BARRON'S

Painless Calculus

Christina Pawlowski-Polanish, M.S.

To my husband, Chris. Thank you for supporting me, believing in me, and being my editor. I love you!

To my mom, Diane, for instilling a love of calculus in me and always teaching me to reach for the stars.

To my brothers, Joseph and Michael, for all of their encouragement along the way.

Published by Kaplan North America, LLC d/b/a Barron's Educational Series
1515 West Cypress Creek Road
Fort Lauderdale, Florida 33309
www.barronseduc.com

ISBN: 978-1-5062-7319-8

10 9 8 7 6 5 4 3 2

Contents

How to Use This Book

Painless calculus? It is not as impossible as you might think. I believe that anyone can learn to love and appreciate one of the most challenging topics in mathematics. I have been teaching math for over 15 years. Math has its own language. Once you understand the language and can visualize the calculus concepts, you too will see that calculus really is painless.

Painless Icons and Features

This book is designed with several unique features to help make learning calculus easy.

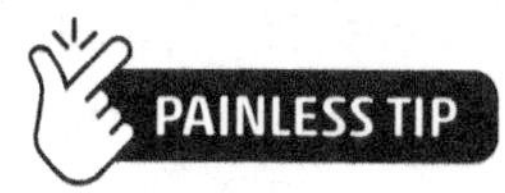

You will see Painless Tips throughout the book. These include helpful tips, hints, and strategies on the surrounding topics.

Caution boxes will help you avoid common pitfalls or mistakes. Be sure to read them carefully.

These boxes translate "math talk" into plain English to make it even easier to understand calculus.

Reminders will call out information that is important to remember. Each reminder will relate to the current chapter or will reference key information you learned in a previous chapter.

BRAIN TICKLERS

There are brain ticklers throughout each chapter in the book. These quizzes are designed to make sure you understand what you've just learned and to test your progress as you move forward in the chapter. Complete all the Brain Ticklers and check your answers. If you get any wrong, make sure to go back and review the topics associated with the questions you missed.

PAINLESS STEPS

Complex procedures are divided into a series of painless steps. These steps help you solve problems in a systematic way. Follow the steps carefully, and you'll be able to solve most calculus problems.

EXAMPLES

Most topics include examples with solutions. If you are having trouble, research shows that writing or copying the problem may help you understand it.

Chapter Breakdown

Chapter One is titled "Limits and Continuity" and serves as an introduction to the building blocks of calculus. Limits allow us to discover one of the major branches of calculus. Limits help us get out of trouble when we are evaluating what seems to be impossible.

Chapter Two discusses one of the major branches of calculus, differentiation. In this chapter, you will learn the different meanings of a derivative and how to use limits to calculate the derivative at a point and the derivative function.

Chapter Three will become your new best friend. In this chapter, you will learn all of the derivative shortcut rules to make finding derivatives faster and easier.

Chapter Four is where you will learn to apply your newfound derivative rules. The derivative has many different meanings and uses. A lot of the math you have learned in the past makes a return, and calculus is there to help explain why it works!

Chapter Five is the other major branch of calculus, antidifferentiation. Think of it as the opposite of differentiation. In this chapter, you will learn that an antiderivative is an integral. Then you will discover all the different integration shortcut techniques.

Chapter Six is much like Chapter Four, where you will gain a deeper understanding of integrals and what they represent. This will get you in tip-top shape for their applications in the next chapter.

Chapter Seven will put your understanding of integrals to the test when it explores the many different applications of antidifferentiation. Many of the three-dimensional formulas you learned in geometry come back to light as this chapter will show you how calculus can find areas and volumes of some pretty wacky shapes. You will also leave with a new appreciation for finding the distance between two points.

Chapter Eight is a great way to tie up loose ends from previous chapters. This is where the connection between derivatives and integrals comes to light. You will discover how well they work together and the amazing connections they have. Be sure to get your pencil and paper ready, and don't be afraid to show off your graphing skills!

If you are learning calculus for the first time or if you are trying to remember what you have learned but may have forgotten, this book is for you. It is a painless introduction to calculus that is both explicit and instructive. Turn forward to the first page. There's nothing to be afraid of. Remember: calculus is painless.

Chapter 1

Limits and Continuity

The concept of a limit plays a vital role in calculus. A limit will help to evaluate expressions that normally would be *undefined* or *indeterminate*.

1+2=3 MATH TALK!

An expression can be undefined in a few ways, such as a 0 in the denominator or approaching $\pm\infty$. Indeterminate is different from undefined. Indeterminate means the expression may still have a value, but alternative methods of solution need to be implemented. Indeterminate forms may be $\frac{0}{0}$, $\frac{\infty}{\infty}$, 1^{∞}, $0 \cdot \infty$, and others.

Definition of a Limit

Given a function $f(x)$, the limit of $f(x)$ as x approaches c is a real number L if $f(x)$ can be made arbitrarily close to L by having x approach close to c (but not equal to c).

This is notated in the table below.

Notation	Read as
$\lim_{x \to c} f(x) = L$	The limit as x approaches c of $f(x)$ equals L

If the limit equals a value L, the limit exists. It is understood that the values of $f(x)$ approach L as x approaches c from the left and from the right.

This is notated in the table below.

Notation	Read as
$\lim_{x \to c^-} f(x) = L$	The limit as x approaches c from the left of $f(x)$ equals L
$\lim_{x \to c^+} f(x) = L$	The limit as x approaches c from the right of $f(x)$ equals L

1+2=3 MATH TALK!

Evaluating a limit from the left means that the x-values are arbitrarily close to c but less than c. Evaluating a limit from the right means that the x-values are arbitrarily close to c but greater than c.

For example, consider the table of values below for $f(x) = x^2$.

x	2.899	2.9	2.99	2.999	3	3.001	3.011	3.11	3.111
$f(x)$	8.404201	8.41	8.9401	8.994001		9.006001	9.066121	9.6721	9.678321

$\lim_{x \to 3^-} f(x) = 9$ $\quad$ $\lim_{x \to 3^+} f(x) = 9$

The $\lim_{x \to 3} x^2 = 9$ because starting on the left of the table and as x moves closer to 3, $f(x)$ is approaching 9. Similarly, starting on the right of the table and as x moves closer to 3, $f(x)$ is approaching 9.

Evaluating limits is *painless*. In addition to looking at a table of values as shown above, another way to evaluate a limit is by looking at the graph of $f(x)$.

Graphic Approach to Limits

If you are given the graph of $f(x)$ or if you have access to a graphing calculator, you can evaluate the limit graphically using a visual approach. Anytime a limit is being evaluated, it must be checked coming from the left and right directions. Only if the left-hand and right-hand limits equal each other does the limit exist.

PAINLESS TIP

When applying the graphic approach to evaluating limits, use your finger to trace along the graph until you arrive at the specific x-value you are trying to approximate. When coming from the left, start all the way at the left end of the graph and move to the right. When coming from the right, start all the way at the right end of the graph and move to the left. If the function values from both sides are equal, the limit exists. Alternatively, if the function values coming from the two sides do not equal each other, the limit does not exist.

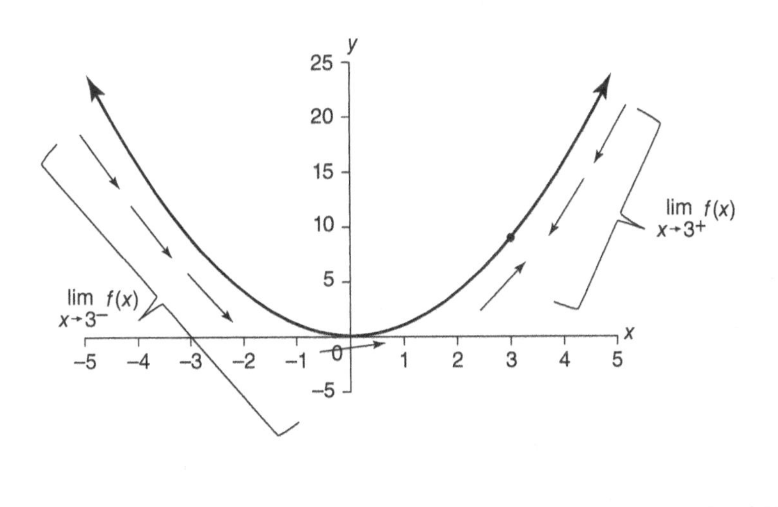

Example 1:

Using the graph below, evaluate the following limits. If the value does not exist, write DNE:

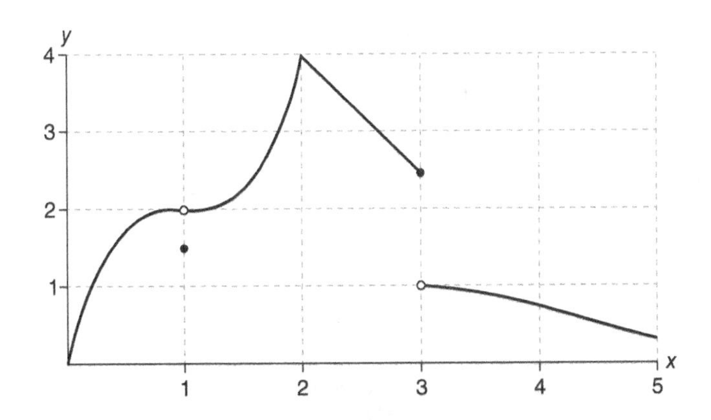

Evaluate	Solution
1. $\lim_{x \to 1^-} f(x)$	2
2. $\lim_{x \to 1^+} f(x)$	2
3. $\lim_{x \to 1} f(x)$	2
4. $\lim_{x \to 2^-} f(x)$	4
5. $\lim_{x \to 2^+} f(x)$	4
6. $\lim_{x \to 2} f(x)$	4
7. $\lim_{x \to 3^-} f(x)$	2.5
8. $\lim_{x \to 3^+} f(x)$	1
9. $\lim_{x \to 3} f(x)$	DNE

CAUTION—Major Mistake Territory!

It is common in the beginning to confuse the concept of a limit with a function value. The $\lim_{x \to a} f(x)$ is not necessarily the same as $f(a)$. When evaluating limits graphically, an open circle at an x-value does not necessarily mean that the limit does not exist even though the function value does not exist. In Example 1, the $\lim_{x \to 1} f(x)$ exists and is equal to 2 even though it is approaching an open circle. However, $f(1) \neq 2$ since there is an open circle there and, instead, $f(1) = 1.5$.

Example 2:

For the function $g(x)$ graphed here, find the following limits or explain why they do not exist.

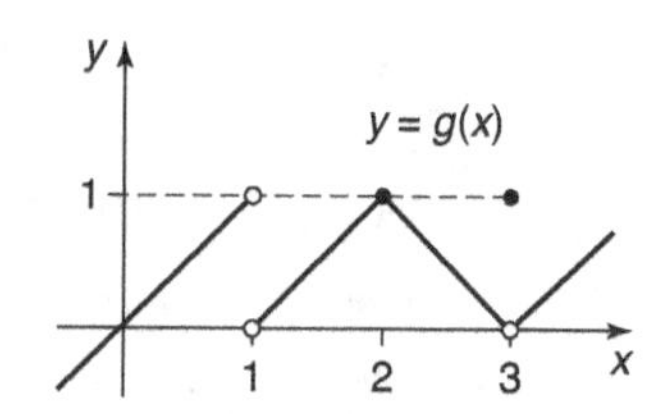

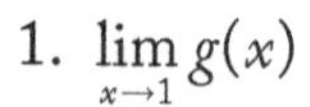

1. $\lim_{x \to 1} g(x)$
2. $\lim_{x \to 2} g(x)$
3. $\lim_{x \to 3} g(x)$

Solution:

1. $\lim_{x \to 1} g(x)$ does not exist because $\lim_{x \to 1^-} g(x) = 1$ and $\lim_{x \to 1^+} g(x) = 0$. Since the left-hand limit does not equal the right-hand limit, the limit does not exist.
2. $\lim_{x \to 2} g(x) = 1$ because $\lim_{x \to 2^-} g(x) = 1$ and $\lim_{x \to 2^+} g(x) = 1$. Since the left-hand limit equals the right-hand limit, the limit does exist and is 1.
3. $\lim_{x \to 3} g(x) = 0$ because $\lim_{x \to 3^-} g(x) = 0$ and $\lim_{x \to 3^+} g(x) = 0$. Since the left-hand limit equals the right-hand limit, the limit does exist and is 0.

Graphic Approach to Limits at Infinity

The symbol ∞ is used to represent *infinity*. Infinity is the idea of something unlimited or without bound, meaning without boundaries. Infinity can appear in limits in two ways: as what x is approaching or what the function approaches.

You are given the graph of $f(x) = \frac{1}{x^2}$. The behavior of the graph is interesting at $x = 0$ since $f(0)$ does not exist. However, by evaluating the limit as x approaches 0, the behavior of the graph can be understood in more detail.

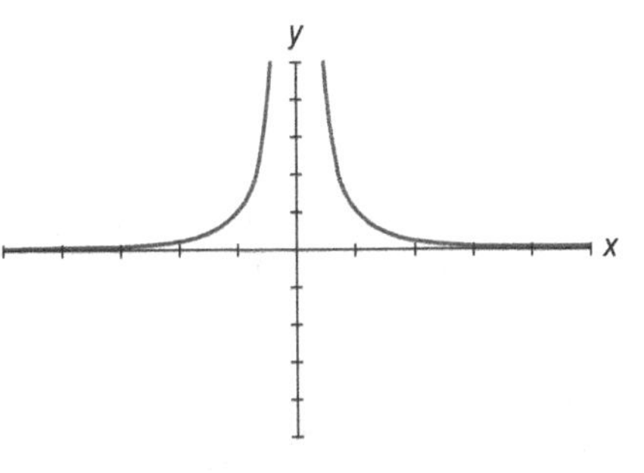

The $\lim_{x \to 0^-} \frac{1}{x^2} = +\infty$. By starting on the left and tracing along the graph until x gets close to zero, our finger moves upward without bound. In other words, $\lim_{x \to 0^-} \frac{1}{x^2} = +\infty$. The same thing happens when starting on the right and tracing along the graph until x gets close to zero: $\lim_{x \to 0^+} \frac{1}{x^2} = +\infty$. Since both limits match, $\lim_{x \to 0} \frac{1}{x^2} = +\infty$. Because the limit approaches positive infinity at this value, the graph has a vertical asymptote at $x = 0$. Since infinity is not a value, $\lim_{x \to 0} \frac{1}{x^2}$ can also be answered as does not exist. However, using infinity is a more specific answer and allows for a better understanding of the graph.

Example 3:

Given the graph of $g(x) = \dfrac{1}{x}$, find the following limits.

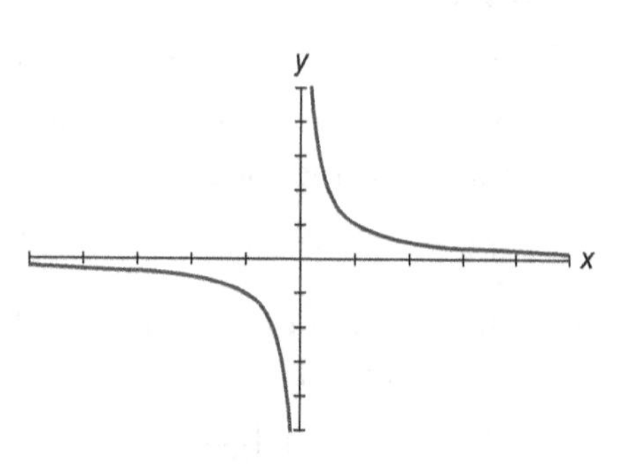

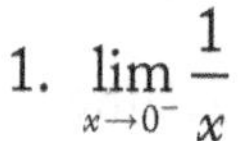

1. $\lim\limits_{x \to 0^-} \dfrac{1}{x}$
2. $\lim\limits_{x \to 0^+} \dfrac{1}{x}$
3. $\lim\limits_{x \to 0} \dfrac{1}{x}$

Solution:

1. $\lim\limits_{x \to 0^-} \dfrac{1}{x} = -\infty$ because starting all the way on the left and tracing the graph until x gets close to zero, our finger moves downward without bound.
2. $\lim\limits_{x \to 0^+} \dfrac{1}{x} = +\infty$ because starting all the way on the right and tracing the graph until x gets close to zero, our finger moves upward without bound.
3. $\lim\limits_{x \to 0} \dfrac{1}{x}$ does not exist because the left-hand limit and the right-hand limit go in two different directions. However, there still exists a vertical asymptote at $x = 0$.

REMINDER

If the $\lim\limits_{x \to a} f(x) = \pm\infty$, there exists a vertical asymptote at $x = a$ and vice versa.

Although infinity is not a real number value, limits can still be evaluated as x approaches either positive or negative infinity.

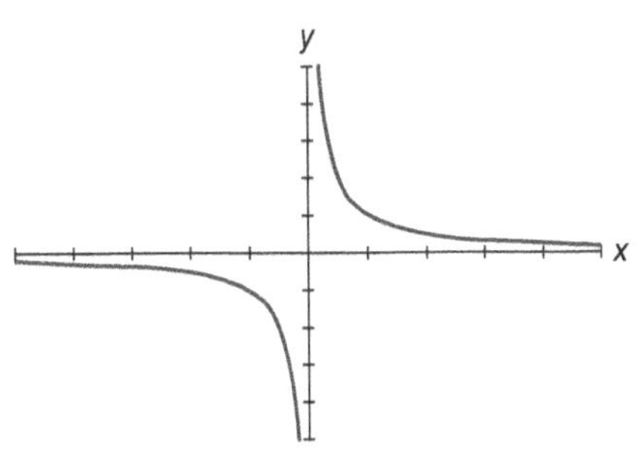

Let's return to the graph of $g(x) = \frac{1}{x}$. As x decreases without bound, $g(x)$ approaches 0. This means that $\lim_{x \to -\infty} \frac{1}{x} = 0$. Similarly, as x increases without bound, $g(x)$ approaches 0. This means that $\lim_{x \to +\infty} \frac{1}{x} = 0$. Graphically, $g(x)$ has a horizontal asymptote at $y = 0$, which is the same value of each of the limits.

Example 4:

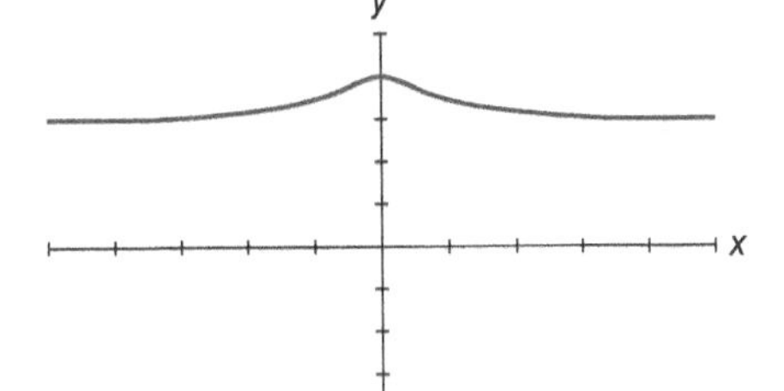

Given the graph of $f(x) = \frac{1}{1+x^2} + 3$, evaluate the following limits.

1. $\lim_{x \to -\infty} f(x)$
2. $\lim_{x \to +\infty} f(x)$
3. What do the above limits imply about the graph of $f(x)$?

Solution:

1. $\lim_{x \to -\infty} f(x) = 3$. By starting in the middle of the graph and tracing along the graph to the left where x decreases without bound, $f(x)$ approaches 3.
2. $\lim_{x \to +\infty} f(x) = 3$. By starting in the middle of the graph and tracing along the graph to the right where x increases without bound, $f(x)$ approaches 3.
3. Since the limits of $f(x)$ as x approaches $\pm\infty$ are 3, there exists a horizontal asymptote at $y = 3$.

REMINDER

If the $\lim_{x \to \pm\infty} f(x) = a$, there exists a horizontal asymptote at $y = a$ and vice versa.

BRAIN TICKLERS Set # 1

Using the graph of $f(x)$, evaluate the following limits.

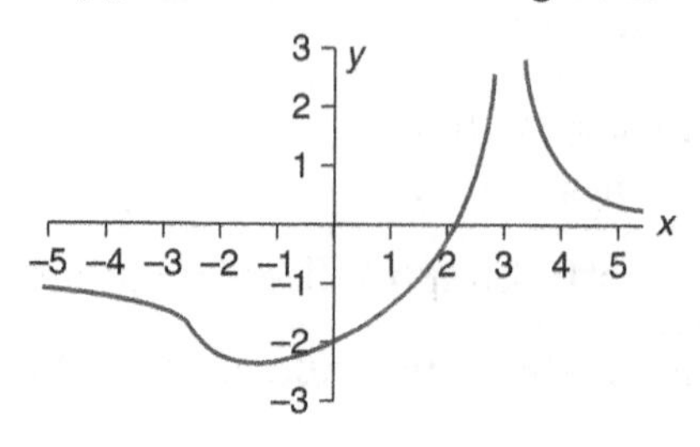

1. $\lim_{x \to 4} f(x)$
2. $\lim_{x \to 3} f(x)$
3. $\lim_{x \to -\infty} f(x)$
4. $\lim_{x \to +\infty} f(x)$

(Answers are on page 34.)

So far, a limit does not exist if it approaches infinity or if the left-hand and right-hand limits are not equal. Additionally, a limit will not exist if the graph is oscillating in such a way that it is impossible to determine the value the function is approaching.

Consider the graph of $f(x) = \sin x$.

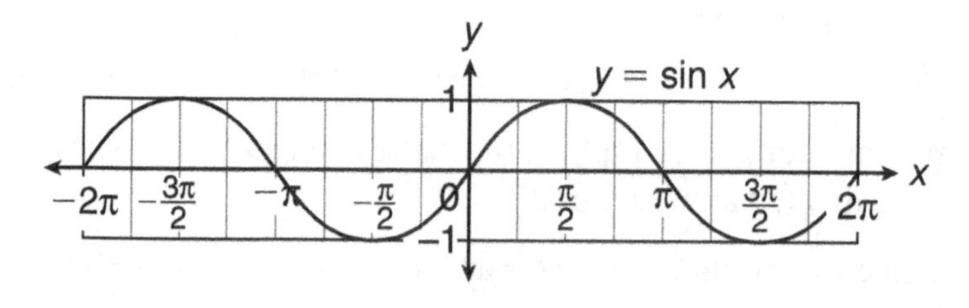

Limits can be evaluated at specific x-values, such as $\lim_{x \to \frac{\pi}{2}} f(x) = 1$ and $\lim_{x \to -2\pi} f(x) = 0$. However, the behavior of the function as x increases or decreases without bound is unable to be determined since the function continuously oscillates between –1 and 1.

Therefore, $\lim_{x \to \pm\infty} f(x)$ does not exist. Another graph to consider is $g(x) = \sin\left(\frac{1}{x}\right)$. The $\lim_{x \to 0} g(x)$ does not exist since the graph is rapidly oscillating as x approaches 0 from both the left and right sides.

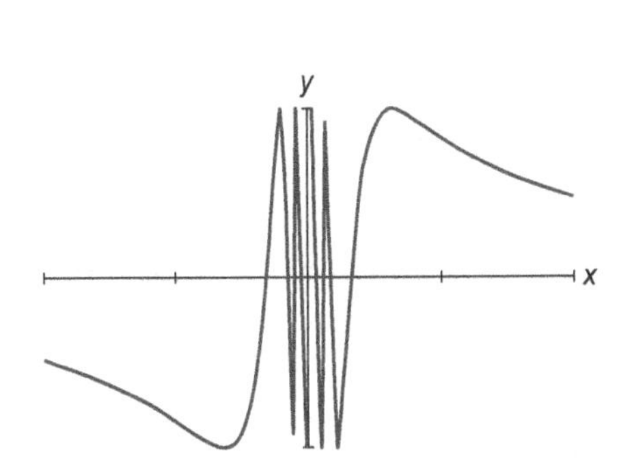

X	Y
-.005	.8733
-.004	.97053
-.003	-.3188
-.002	.46777
-.001	-.8269
0	ERROR
.001	.82688
.002	-.4678
.003	.31885
.004	-.9705
.005	-.8733

Look at the table of values for $g(x)$. As x gets closer to 0 from both the left and right sides, the function values are oscillating from negative to positive so quickly that no common y-value is being approached.

There are three cases where a limit does not exist:

1. The left-hand limit does not equal the right-hand limit: $\lim_{x \to a^-} f(x) \neq \lim_{x \to a^+} f(x)$.
2. The limit approaches positive or negative infinity: $\lim_{x \to a} f(x) = \pm\infty$.
3. The graph of the function is oscillating or periodic.

Properties of Limits

There are nine limit properties to learn and apply when evaluating limits algebraically. Let a, k, and n represent real numbers.

Property 1: The limit of a constant, k, is equal to that constant.

$$\lim_{x \to a} k = k$$

Property 2: The limit of the identity function, $f(x) = x$, is equal to the x-value the limit is approaching.

$$\lim_{x \to a} x = a$$

Property 3: The limit of the reciprocal function, $g(x) = \frac{1}{x}$, as x decreases or increases without bound is equal to 0.

$$\lim_{x \to -\infty} \frac{1}{x} = 0 \quad \text{or} \quad \lim_{x \to +\infty} \frac{1}{x} = 0$$

Property 4: The limit of the sum or difference of two or more functions is equal to the sum or difference of the limits of the functions.

$$\lim_{x \to a} \left[f(x) \pm g(x) \right] = \lim_{x \to a} f(x) \pm \lim_{x \to a} g(x)$$

Property 5: The limit of the product of two or more functions is equal to the product of the limits of the functions.

$$\lim_{x \to a} \left[f(x) \bullet g(x) \right] = \lim_{x \to a} f(x) \bullet \lim_{x \to a} g(x)$$

Property 6: The limit of the quotient of two functions is equal to the quotient of the limits of the functions if the limit of the denominator is not equal to zero.

$$\lim_{x \to a} \left[\frac{f(x)}{g(x)} \right] = \frac{\lim_{x \to a} f(x)}{\lim_{x \to a} g(x)}, \ \lim_{x \to a} g(x) \neq 0$$

Property 7: The limit of a power of a function is equal to the power of the limit.

$$\lim_{x \to a} ((f(x))^n) = \left(\lim_{x \to a} f(x) \right)^n$$

Property 8: The limit of a root of a function is equal to the root of the limit provided that the limit of the function is nonnegative if the root is even.

$$\lim_{x \to a} \sqrt[n]{f(x)} = \sqrt[n]{\lim_{x \to a} f(x)}, \ \lim_{x \to a} f(x) > 0 \text{ if } n \text{ is even}$$

Property 9: The limit of a constant, k, multiplied by a function is equal to the constant multiplied by the limit of the function.

$$\lim_{x \to a} kf(x) = k \bullet \lim_{x \to a} f(x)$$

Example 5:

If $\lim_{x \to 2} f(x) = 1$ and $\lim_{x \to 2} g(x) = 5$, evaluate the following.

1. $\lim_{x \to 2}(7 + g(x))$
2. $\lim_{x \to 2}\left(\frac{f(x)}{x}\right)$

Solution:

1. $\lim_{x \to 2}(7 + g(x)) = \lim_{x \to 2} 7 + \lim_{x \to 2} g(x)$ by Property 4

   $= 7 + 5$ by Property 1 and by substituting the given information

   $= 12.$

2. $\lim_{x \to 2}\left(\frac{f(x)}{x}\right) = \frac{\lim_{x \to 2} f(x)}{\lim_{x \to 2} x}$ by Property 6

   $= \frac{1}{2}$ by substituting the given information and by Property 2.

Example 6:

Using limit properties, evaluate $\lim_{x \to 5}(x^2 - 4x + 3)$.

Solution:

Using Property 4, find the limit of each of the terms in the sum and difference.

$$\lim_{x \to 5}(x^2) - \lim_{x \to 5}(4x) + \lim_{x \to 5} 3$$

Rewrite by applying Property 7 and Property 9.

$$\left(\lim_{x \to 5} x\right)^2 - 4 \bullet \lim_{x \to 5} x + \lim_{x \to 5} 3$$

Evaluate the limit using Property 1 and Property 2.

$$(5)^2 - 4(5) + 3 = 8$$

As shown in Example 6, the limit of any polynomial function is equal to the function evaluated at the x-value. If

$f(x) = x^2 - 4x + 3$, then $f(5) = (5)^2 - 4(5) + 3 = 8$. This is the same value as $\lim_{x \to 5}(x^2 - 4x + 3)$. This would also be true for rational functions.

1+2=3 MATH TALK!

If $f(x)$ is a polynomial function, $\lim_{x \to a} f(x) = f(a)$.

In other words, when evaluating the limit of a polynomial function, substitute the x-value into the function and simplify.

Similarly, if a rational function, $h(x)$, is the ratio of two polynomial functions, $f(x)$ and $g(x)$, $\lim_{x \to a} h(x) = \frac{f(a)}{g(a)}$ as long as $g(a) \neq 0$.

In other words, when evaluating the limit of a rational function, substitute the x-value into the functions of the numerator and denominator and then simplify.

Example 7:

Evaluate $\lim_{t \to 0} \frac{6t - 9}{t^3 - 12t + 3}$.

Solution:

Since the limit is of a rational function, substitute 0 for t into the numerator and denominator and then simplify:

$$\lim_{t \to 0} \frac{6t - 9}{t^3 - 12t + 3} = \frac{6(0) - 9}{(0)^3 - 12(0) + 3} = \frac{-9}{3} = -3$$

BRAIN TICKLERS Set # 2

Evaluate the following limits.

1. $\lim_{x \to 3} -2$

2. $\lim_{x \to 3} x$

3. $\lim_{x \to 3} 5x^3 - 10x + 7$

4. $\lim_{x \to 3} \frac{x^2 - 9}{x + 1}$

(Answers are on page 34.)

Algebraic Approach to Limits

Evaluating limits algebraically is *painless*. There are three steps to evaluating the limit of a function algebraically.

Step 1: Separate the limit into the left-hand and right-hand limits if necessary.

Step 2: Apply the appropriate limit properties.

Step 3: Simplify.

Piecewise functions are different from polynomial and rational functions. It is important to consider the left-hand and right-hand limits for piecewise functions.

Example 8:

Given $f(x) = \begin{cases} \dfrac{1}{x+2}, x < -2 \\ x^2 - 5, -2 \leq x \leq 3 \\ \sqrt{x+13}, x > 3 \end{cases}$, evaluate the following.

1. $\lim\limits_{x \to -2} f(x)$
2. $\lim\limits_{x \to 0} f(x)$
3. $\lim\limits_{x \to 3} f(x)$

Solution:

1. Since $\lim\limits_{x \to -2} f(x)$ has x approaching –2, which is on the boundary of two different functions, it is necessary to split the limit into the left-hand and right-hand limits and to use the appropriate functions for each side.

Left-Hand Limit	Right-Hand Limit
$\lim\limits_{x \to -2^-} f(x) = \lim\limits_{x \to -2^-} \dfrac{1}{x+2}$	$\lim\limits_{x \to -2^+} f(x) = \lim\limits_{x \to -2^+} x^2 - 5$
$= \dfrac{1}{-2+2}$	$= (-2)^2 - 5$
$= \dfrac{1}{0}$	$= 4 - 5$
The limit does not exist.	$= -1$

Since $\lim\limits_{x \to -2^-} f(x)$ does not exist, $\lim\limits_{x \to -2} f(x)$ does not exist.

2. Since $\lim_{x\to 0} f(x)$ has x approaching 0, which is in the interval of the middle polynomial function, it is not necessary to split the limit into the left-hand and right-hand limits.

   The $\lim_{x\to 0} f(x) = \lim_{x\to 0} x^2 - 5 = (0)^2 - 5 = -5$.

3. Since $\lim_{x\to 3} f(x)$ has x approaching 3, which is on the boundary of two different functions, it is necessary to split the limit into the left-hand and right-hand limits and to use the appropriate functions for each side.

Left-Hand Limit	Right-Hand Limit
$\lim_{x\to 3^-} f(x) = \lim_{x\to 3^-} x^2 - 5$	$\lim_{x\to 3^+} f(x) = \lim_{x\to 3^+} \sqrt{x+13}$
$= (3)^2 - 5$	$= \sqrt{3+13}$
$= 9 - 5$	$= \sqrt{16}$
$= 4$	$= 4$

Since $\lim_{x\to 3^-} f(x) = \lim_{x\to 3^+} f(x)$, then $\lim_{x\to 3} f(x) = 4$.

The limits can be verified by using the graph of the piecewise function as well.

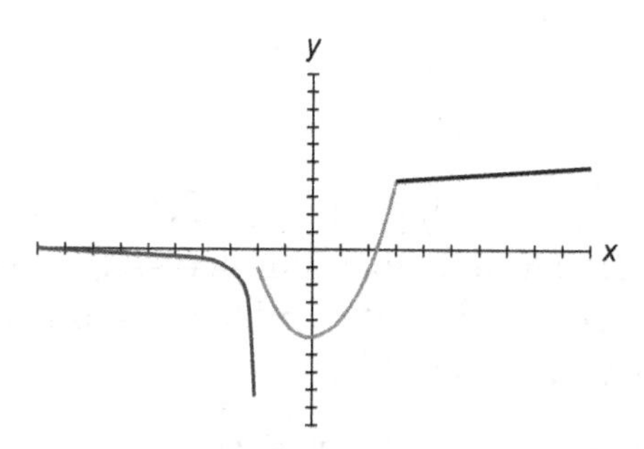

Sometimes when limits are evaluated algebraically, the answers are indeterminate. These can take on various forms, such as $\frac{0}{0}$. If this occurs, different algebraic techniques need to be used in order to determine if the limit exists.

CAUTION—Major Mistake Territory!

If a limit simplifies to $\frac{0}{0}$, it is not equivalent to 0 or 1. This is indeterminant, meaning that the limit may or may not exist and a different algebraic technique must be used to evaluate the limit.

If evaluating a limit of a rational function yields an indeterminate form, try factoring the numerator and denominator and remove any common factors. Then evaluate the limit again.

Example 9:

Evaluate $\lim\limits_{x \to 3} \frac{x^2 - 9}{x - 3}$.

Solution:

$$\lim_{x \to 3} \frac{x^2 - 9}{x - 3} = \frac{\lim\limits_{x \to 3} x^2 - 9}{\lim\limits_{x \to 3} x - 3} = \frac{(3)^2 - 9}{(3) - 3} = \frac{9 - 9}{3 - 3} = \frac{0}{0}$$

This is indeterminate. Since it is a rational function, an alternative method is to factor the numerator and denominator to see if any common factors can be removed.

$$\lim_{x \to 3} \frac{x^2 - 9}{x - 3} = \lim_{x \to 3} \frac{(x - 3)(x + 3)}{x - 3}$$

By removing the common factor $(x - 3)$, the limit simplifies to $\lim\limits_{x \to 3}(x + 3) = 3 + 3 = 6$.

Example 10:

Evaluate $\lim\limits_{x \to 5} \frac{x^2 - 3x - 10}{x^2 - 10x + 25}$.

Solution:

$$\lim_{x \to 5} \frac{x^2 - 3x - 10}{x^2 - 10x + 25} = \frac{\lim\limits_{x \to 5} x^2 - 3x - 10}{\lim\limits_{x \to 5} x^2 - 10x + 25} = \frac{(5)^2 - 3(5) - 10}{(5)^2 - 10(5) + 25} = \frac{0}{0}$$

This is indeterminate. Since it is a rational function, an alternative method is to factor the numerator and denominator to see if any common factors can be removed.

$$\lim_{x\to5}\frac{x^2-3x-10}{x^2-10x+25}=\lim_{x\to5}\frac{(x-5)(x+2)}{(x-5)(x-5)}$$

By removing the common factor $(x-5)$, the limit simplifies to $\lim_{x\to5}\frac{x+2}{x-5}=\frac{\lim_{x\to5}x+2}{\lim_{x\to5}x-5}=\frac{7}{0}$. So, the limit does not exist.

BRAIN TICKLERS Set # 3

1. If $f(x)=\begin{cases}x-1,\ x\le3\\3x-7,\ x>3\end{cases}$, evaluate $\lim_{x\to3}f(x)$.
2. Evaluate $\lim_{x\to3}\frac{x^2-2x}{x+1}$.
3. Evaluate $\lim_{x\to4}\frac{x^2-16}{x-4}$.
4. Evaluate $\lim_{x\to1^+}\frac{x^4-1}{x-1}$.

(Answers are on page 34.)

Finding the limits of radical expressions involves the use of Property 8.

Example 11:

Evaluate $\lim_{x\to1}\sqrt{2x+4}$.

Solution:

Use Property 8.

$$\lim_{x\to1}\sqrt{2x+4}=\sqrt{\lim_{x\to1}(2x+4)}=\sqrt{2(1)+4}=\sqrt{6}$$

Sometimes rational expressions involve radicals. When the limit is indeterminate, another technique must be used. For instance, you can often multiply the rational expression by the conjugate of either the numerator or denominator, whichever one has the radical in the expression.

Example 12:

Evaluate $\lim\limits_{x \to 1} \dfrac{x-1}{\sqrt{x}-1}$.

Solution:

$$\lim_{x \to 1} \frac{x-1}{\sqrt{x}-1} = \frac{\lim\limits_{x \to 1} x - 1}{\lim\limits_{x \to 1} \sqrt{x} - 1} = \frac{1-1}{\sqrt{1}-1} = \frac{0}{0}$$

This is indeterminate. Since this is a rational expression with a radical binomial in the denominator, multiply the numerator and denominator by the conjugate of the denominator. After multiplying, remove any common factors.

$$\lim_{x \to 1} \frac{x-1}{\sqrt{x}-1} = \lim_{x \to 1} \frac{x-1}{\sqrt{x}-1} \bullet \frac{\sqrt{x}+1}{\sqrt{x}+1} = \lim_{x \to 1} \frac{(x-1)(\sqrt{x}+1)}{x-1}$$

After removing the common factor $(x - 1)$, the limit simplifies to $\lim\limits_{x \to 1} \sqrt{x} + 1 = \sqrt{1} + 1 = 2$.

PAINLESS TIP

When multiplying by the conjugate, it is necessary to multiply only the conjugate expressions together. In Example 12, only the conjugates were multiplied together and not the terms in the numerator. Leaving the numerator as a product made it easier to see the common factors.

Example 13:

Find $\lim\limits_{x \to -1} \dfrac{\sqrt{x+5}-2}{x+1}$.

Solution:

$$\lim_{x \to -1} \frac{\sqrt{x+5}-2}{x+1} = \frac{\lim\limits_{x \to -1} \sqrt{x+5} - 2}{\lim\limits_{x \to -1} x + 1} = \frac{\sqrt{-1+5}-2}{-1+1} = \frac{0}{0}$$

This is indeterminate. Since this is a rational expression with a radical binomial in the numerator, multiply the numerator and denominator by the conjugate of the numerator. After multiplying, remove any common factors.

$$\lim_{x\to-1}\frac{\sqrt{x+5}-2}{x+1}=\lim_{x\to-1}\frac{\sqrt{x+5}-2}{x+1}\bullet\frac{\sqrt{x+5}+2}{\sqrt{x+5}+2}$$
$$=\lim_{x\to-1}\frac{x+5-4}{(x+1)(\sqrt{x+5}+2)}=\lim_{x\to-1}\frac{x+1}{(x+1)(\sqrt{x+5}+2)}$$

After removing the common factor $(x + 1)$, the limit simplifies to $\lim_{x\to-1}\frac{1}{\sqrt{x+5}+2}=\frac{1}{\sqrt{(-1)+5}+2}=\frac{1}{\sqrt{4}+2}=\frac{1}{4}$.

CAUTION—Major Mistake Territory!

Whether the radical binomial expression is in the denominator or is in the numerator does not matter. In the past, multiplying conjugates was to rationalize the denominator of an expression. When evaluating limits, the radical binomial can appear in either the numerator or denominator.

Limits for other functions like *exponential*, *logarithmic*, and *trigonometric* follow the limit properties as well and can often be evaluated simply by substituting in the value x is approaching. Graphing the function to evaluate the limit is also a good way to verify answers.

Example 14:

Evaluate the following limits.

1. $\lim_{x\to\pi}\frac{1}{2}\sin\left(\frac{x}{6}\right)$
2. $\lim_{x\to-10} e^x$
3. $\lim_{x\to1}\ln(x)$

Solution:

1. $\lim_{x\to\pi}\frac{1}{2}\sin\left(\frac{x}{6}\right)=\frac{1}{2}\bullet\lim_{x\to\pi}\sin\left(\frac{x}{6}\right)=\frac{1}{2}\bullet\sin\left(\frac{\pi}{6}\right)=\frac{1}{2}\bullet\frac{1}{2}=\frac{1}{4}$

2. $\lim_{x \to -10} e^{x} = e^{-10} = \frac{1}{e^{10}}$
3. $\lim_{x \to 1} \ln(x) = \ln(1) = 0$

Two important trigonometric limits help to evaluate limits algebraically:

1. $\lim_{x \to 0} \frac{\sin x}{x} = 1$ and $\lim_{x \to 0} \frac{x}{\sin x} = 1$
2. $\lim_{x \to 0} \frac{\cos x - 1}{x} = 0$ and $\lim_{x \to 0} \frac{1 - \cos x}{x} = 0$

1+2=3 MATH TALK!

The first trigonometric limit can further extend to the following: $\lim_{x \to 0} \frac{\sin(ax)}{x} = a$, where a is a constant not equal to zero.

Example 15:

Evaluate the following limits:

1. $\lim_{x \to 0} \frac{\sin x + x}{x}$
2. $\lim_{x \to 0} \frac{\sin(12x)}{x}$
3. $\lim_{x \to 0} \frac{\tan x}{x}$
4. $\lim_{x \to 0} \frac{\sin(2x)}{\sin(5x)}$

Solution:

1. $\lim_{x \to 0} \frac{\sin x + x}{x} = \lim_{x \to 0} \left(\frac{\sin x}{x} + \frac{x}{x} \right)$ by expanding the fraction. Use Property 4.

$$\lim_{x \to 0} \left(\frac{\sin x}{x} + \frac{x}{x} \right) = \lim_{x \to 0} \frac{\sin x}{x} + \lim_{x \to 0} 1 = 1 + 1 = 2$$

2. $\lim_{x \to 0} \frac{\sin(12x)}{x} = 12$

3. $\lim_{x\to 0}\frac{\tan x}{x} = \frac{\tan(0)}{0} = \frac{0}{0}$, which is indeterminate. Since $\tan x = \frac{\sin x}{\cos x}$, use substitution.

$$\lim_{x\to 0}\frac{\tan x}{x} = \lim_{x\to 0}\frac{\frac{\sin x}{\cos x}}{x} = \lim_{x\to 0}\frac{\sin x}{x \bullet \cos x}$$

$$= \lim_{x\to 0}\frac{\sin x}{x} \bullet \lim_{x\to 0}\frac{1}{\cos x} = 1 \bullet 1 = 1$$

4. $\lim_{x\to 0}\frac{\sin(2x)}{\sin(5x)}$ looks like the trigonometric limit $\lim_{x\to 0}\frac{\sin(ax)}{x} = a$.
Divide the numerator and denominator by x.

$$\lim_{x\to 0}\frac{\sin(2x)}{\sin(5x)} = \lim_{x\to 0}\frac{\frac{\sin(2x)}{x}}{\frac{\sin(5x)}{x}} = \frac{\lim_{x\to 0}\frac{\sin(2x)}{x}}{\lim_{x\to 0}\frac{\sin(5x)}{x}} = \frac{2}{5}$$

BRAIN TICKLERS Set # 4

Evaluate each of the following limits.

1. $\lim_{x\to 0}\frac{\sqrt{x+4}-2}{x}$

2. $\lim_{x\to \frac{\pi}{2}} 2\tan\left(\frac{x}{2}\right)$

3. $\lim_{x\to 0}\frac{1-\cos x}{x}$

4. $\lim_{x\to 0}\frac{2x+\sin x}{x}$

(Answers are on page 34.)

Algebraic Approach to Limits at Infinity

Limits approaching positive or negative infinity and limits that result in infinity can also be found through algebraic means.

When taking the limit as x approaches $\pm\infty$ of a polynomial function, the end behavior of the function is being evaluated. No matter how many terms the polynomial function has, the end behavior will depend on the leading coefficient of the term with the highest power, or degree. To evaluate the limit of a polynomial function as x approaches either $\pm\infty$, determine if the leading coefficient of the polynomial is either positive or negative and determine if the degree is either even or odd. This leads to four possible scenarios as explained in the table below.

Leading Coefficient	Degree	End Behavior
Positive	Even	Graph rises on the left and right ↖ ↗
Positive	Odd	Graph falls on the left and rises on the right ↙ ↗
Negative	Even	Graph falls on the left and right ↙ ↘
Negative	Odd	Graph rises on the left and falls on the right ↖ ↘

This behavior can easily be identified in two of our parent functions, $f(x) = x^2$ and $g(x) = x^3$.

For $f(x) = x^2$, the leading coefficient (1) is positive and the degree (2) is even. As seen in the accompanying figure, $\lim\limits_{x \to \pm\infty} x^2 = +\infty$.

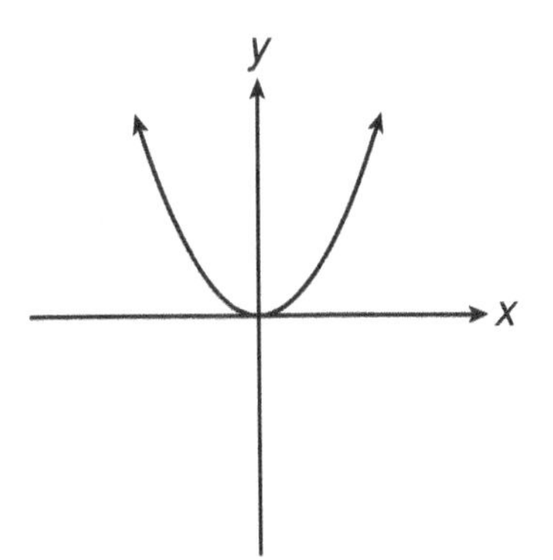

For $g(x) = x^3$, the leading coefficient (1) is positive and the degree (3) is odd. As seen in the accompanying figure, $\lim\limits_{x \to -\infty} x^3 = -\infty$ and $\lim\limits_{x \to +\infty} x^3 = +\infty$.

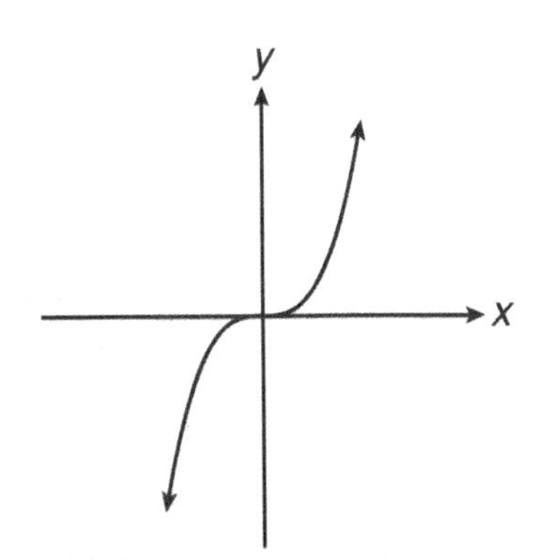

Example 16:

Given $p(x) = -2x^3 - 7x^2 - 4x + 6$, evaluate $\lim_{x \to -\infty} p(x)$ and $\lim_{x \to +\infty} p(x)$. Confirm your answer graphically.

Solution:

When evaluating limits for a polynomial function as x approaches $\pm\infty$, refer to the function's end behavior. Since the leading coefficient of $p(x)$ is –2, which is negative, and since the degree of $p(x)$ is 3, which is odd, the graph will rise on the left and fall on the right.

This means $\lim_{x \to -\infty} p(x) = +\infty$ and $\lim_{x \to +\infty} p(x) = -\infty$.

Confirm graphically:

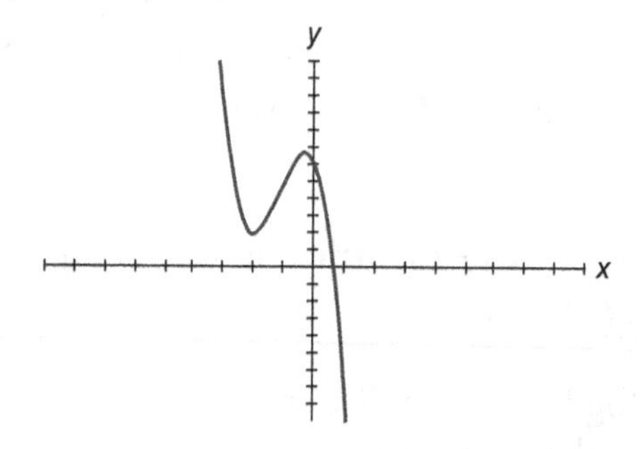

Rational functions have their own type of infinite limit behavior, which is illustrated in the table below.

$f(x)$	Graph	$x \to \pm\infty$	$x \to a$
$\dfrac{1}{x-a}$	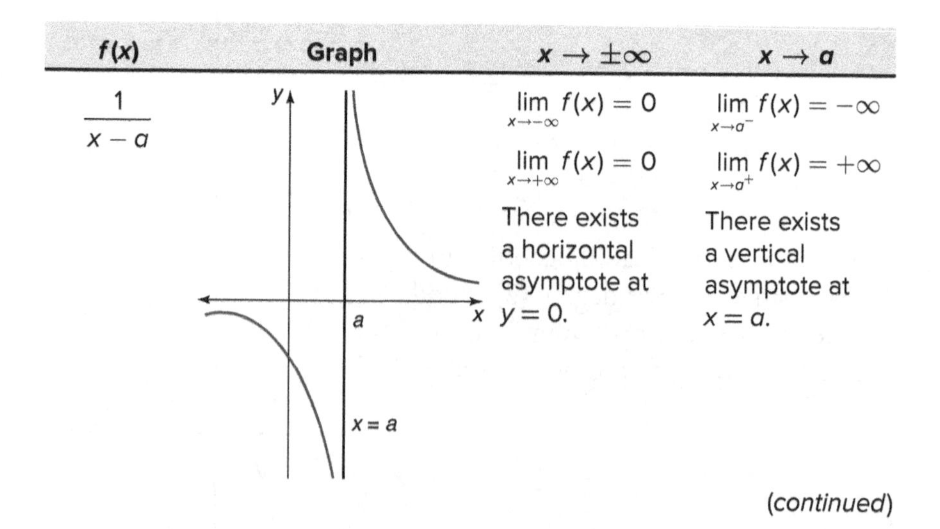	$\lim_{x \to -\infty} f(x) = 0$ $\lim_{x \to +\infty} f(x) = 0$ There exists a horizontal asymptote at $y = 0$.	$\lim_{x \to a^-} f(x) = -\infty$ $\lim_{x \to a^+} f(x) = +\infty$ There exists a vertical asymptote at $x = a$.

(continued)

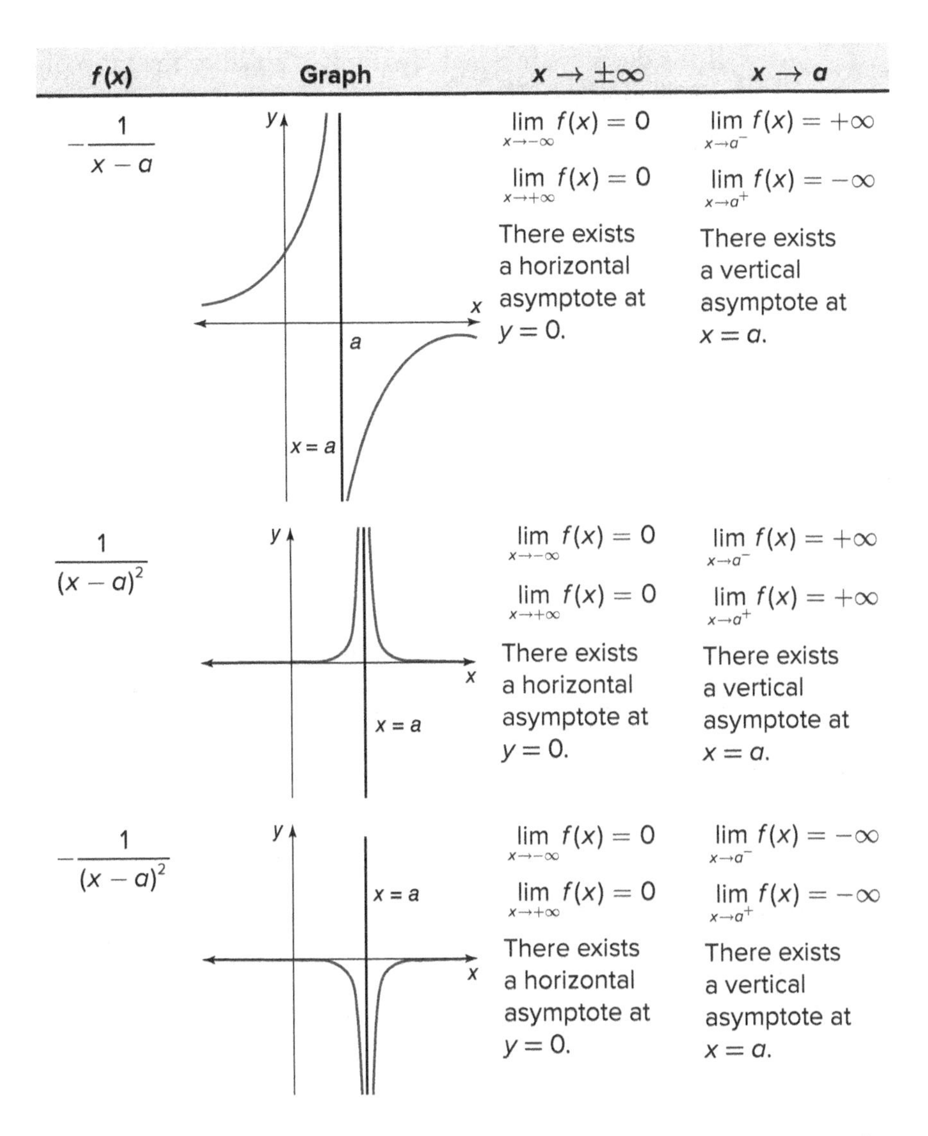

$f(x)$	Graph	$x \to \pm\infty$	$x \to a$
$-\dfrac{1}{x-a}$		$\lim_{x\to-\infty} f(x) = 0$ $\lim_{x\to+\infty} f(x) = 0$ There exists a horizontal asymptote at $y = 0$.	$\lim_{x\to a^-} f(x) = +\infty$ $\lim_{x\to a^+} f(x) = -\infty$ There exists a vertical asymptote at $x = a$.
$\dfrac{1}{(x-a)^2}$		$\lim_{x\to-\infty} f(x) = 0$ $\lim_{x\to+\infty} f(x) = 0$ There exists a horizontal asymptote at $y = 0$.	$\lim_{x\to a^-} f(x) = +\infty$ $\lim_{x\to a^+} f(x) = +\infty$ There exists a vertical asymptote at $x = a$.
$-\dfrac{1}{(x-a)^2}$		$\lim_{x\to-\infty} f(x) = 0$ $\lim_{x\to+\infty} f(x) = 0$ There exists a horizontal asymptote at $y = 0$.	$\lim_{x\to a^-} f(x) = -\infty$ $\lim_{x\to a^+} f(x) = -\infty$ There exists a vertical asymptote at $x = a$.

Not all rational functions will have constants in the numerator and, instead, are a ratio of polynomial functions. If this is the case, evaluating rational functions as x approaches infinity is *painless*. Follow these two steps.

Step 1: Identify the degree of the numerator and denominator.

Step 2: Compare the degrees.

- If the degree in the numerator is less than the degree in the denominator, the limit is equal to 0.

- If the degrees are equal, the limit is equal to the ratio of the leading coefficients.
- If the degree in the numerator is greater than the degree in the denominator, the limit does not exist.

Example 17:

Evaluate the following limits.

1. $\lim_{x \to -\infty} \frac{4x^2 - x}{2x^3 - 5}$
2. $\lim_{x \to \infty} \frac{3x + 5}{6x - 8}$
3. $\lim_{x \to \infty} \frac{5x^3 - 2x^2 + 1}{1 - 3x}$

Solution:

1. For $\lim_{x \to -\infty} \frac{4x^2 - x}{2x^3 - 5}$, the degree of the numerator is 2 and the degree of the denominator is 3. Since $2 < 3$, $\lim_{x \to -\infty} \frac{4x^2 - x}{2x^3 - 5} = 0$.
2. For $\lim_{x \to \infty} \frac{3x + 5}{6x - 8}$, the degree of the numerator is 1 and the degree of the denominator is 1. Since $1 = 1$, $\lim_{x \to \infty} \frac{3x + 5}{6x - 8} = \frac{3}{6} = \frac{1}{2}$.
3. For $\lim_{x \to \infty} \frac{5x^3 - 2x^2 + 1}{1 - 3x}$, the degree of the numerator is 3 and the degree of the denominator is 1. Since $3 > 1$, $\lim_{x \to \infty} \frac{5x^3 - 2x^2 + 1}{1 - 3x}$ does not exist or, more specifically, approaches $-\infty$.

Example 18:

Evaluate $\lim_{x \to \infty} \sqrt[3]{\frac{4x^2 + 5}{6x^2 - 8}}$.

Solution:

Using Property 8, $\lim_{x \to \infty} \sqrt[3]{\frac{4x^2 + 5}{6x^2 - 8}} = \sqrt[3]{\lim_{x \to \infty} \frac{4x^2 + 5}{6x^2 - 8}}$. Since the degrees of the numerator and denominator both equal 2,

$$\sqrt[3]{\lim_{x \to \infty} \frac{4x^2 + 5}{6x^2 - 8}} = \sqrt[3]{\frac{4}{6}} = \sqrt[3]{\frac{2}{3}}.$$

BRAIN TICKLERS Set # 5

Evaluate each of the following limits.

1. $\lim_{x \to -\infty} (7x^5 - 4x^3 + 2x - 9)$
2. $\lim_{x \to +\infty} (-4x^8 + 17x^3 - 5x + 1)$
3. $\lim_{x \to +\infty} \dfrac{4x^2 + 25}{2x^3 - 60}$
4. $\lim_{x \to -\infty} \dfrac{3x^5 - 21}{5x - 1}$

(Answers are on page 34.)

Exponential and logarithmic functions have their own infinite limit behaviors as shown in the table below.

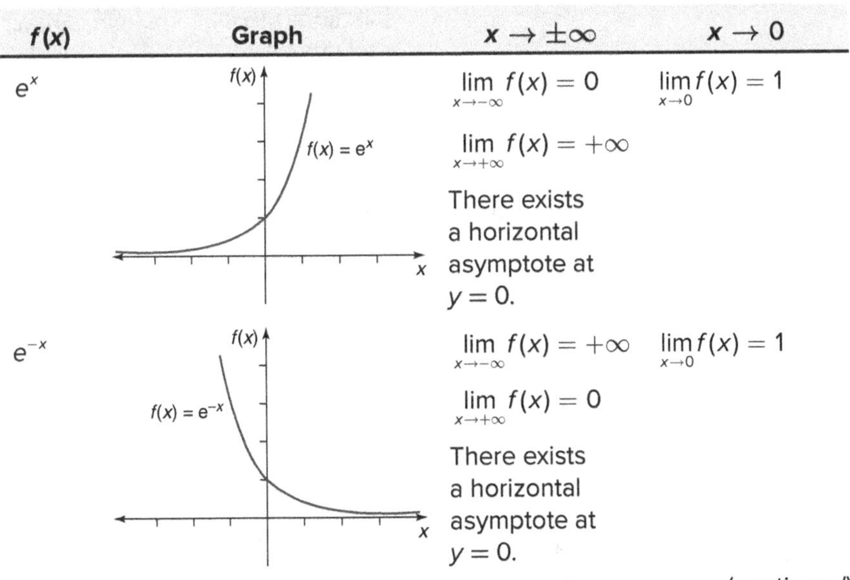

$f(x)$	Graph	$x \to \pm\infty$	$x \to 0$
e^x		$\lim_{x \to -\infty} f(x) = 0$ $\lim_{x \to +\infty} f(x) = +\infty$ There exists a horizontal asymptote at $y = 0$.	$\lim_{x \to 0} f(x) = 1$
e^{-x}		$\lim_{x \to -\infty} f(x) = +\infty$ $\lim_{x \to +\infty} f(x) = 0$ There exists a horizontal asymptote at $y = 0$.	$\lim_{x \to 0} f(x) = 1$

(continued)

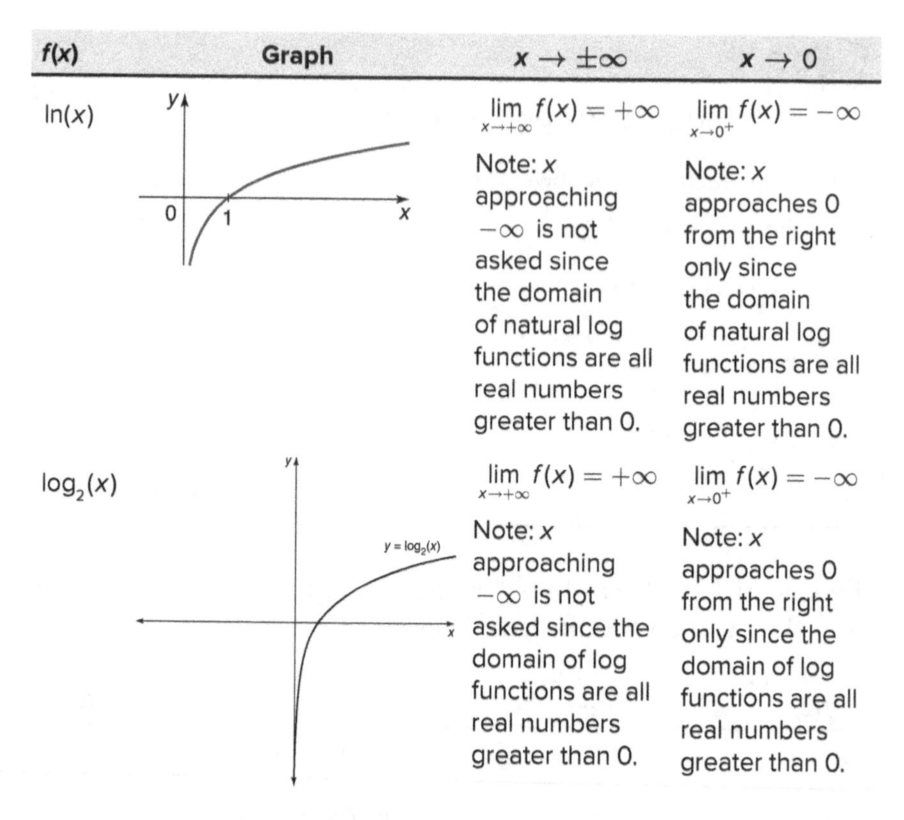

$f(x)$	Graph	$x \to \pm\infty$	$x \to 0$
$\ln(x)$		$\lim_{x \to +\infty} f(x) = +\infty$ Note: x approaching $-\infty$ is not asked since the domain of natural log functions are all real numbers greater than 0.	$\lim_{x \to 0^+} f(x) = -\infty$ Note: x approaches 0 from the right only since the domain of natural log functions are all real numbers greater than 0.
$\log_2(x)$		$\lim_{x \to +\infty} f(x) = +\infty$ Note: x approaching $-\infty$ is not asked since the domain of log functions are all real numbers greater than 0.	$\lim_{x \to 0^+} f(x) = -\infty$ Note: x approaches 0 from the right only since the domain of log functions are all real numbers greater than 0.

Another interesting function is the absolute value function, $f(x) = |x|$. The absolute value function can also be written as a piecewise function:

$$f(x) = |x| = \begin{cases} -x, \ x < 0 \\ 0, \ x = 0 \\ x, \ x > 0 \end{cases}$$

1+2=3 MATH TALK!

Think of the absolute value of x as two different pieces of two different functions. On the left it behaves like $f(x) = -x$, and on the right it behaves like $f(x) = +x$.

Absolute value functions have their own infinite limit behaviors as shown in the table below.

$f(x)$	Graph	$x \to \pm\infty$	$x \to 0$
$\lvert x \rvert$	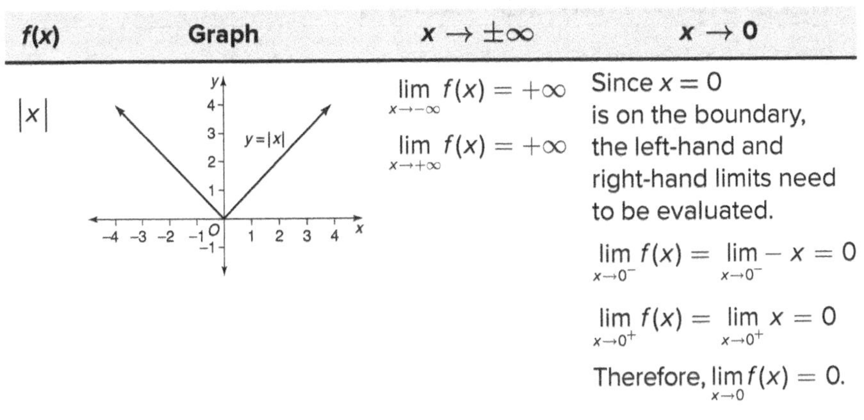	$\lim_{x \to -\infty} f(x) = +\infty$ $\lim_{x \to +\infty} f(x) = +\infty$	Since $x = 0$ is on the boundary, the left-hand and right-hand limits need to be evaluated. $\lim_{x \to 0^-} f(x) = \lim_{x \to 0^-} -x = 0$ $\lim_{x \to 0^+} f(x) = \lim_{x \to 0^+} x = 0$ Therefore, $\lim_{x \to 0} f(x) = 0$.

Example 19:

If $f(x) = \frac{|x|}{x}$, find $\lim_{x \to 0} f(x)$.

Solution:

Rewrite $f(x)$ as a piecewise function.

$$f(x) = \begin{cases} \frac{-x}{x}, x < 0 \\ \frac{+x}{x}, x > 0 \end{cases}$$

This simplifies to $f(x) = \begin{cases} -1, x < 0 \\ 1, x > 0 \end{cases}$. At $x = 0$, the graph is undefined.

To evaluate $\lim_{x \to 0} f(x)$, consider the left-hand and right-hand limits. The $\lim_{x \to 0^-} f(x) = -1$ and $\lim_{x \to 0^+} f(x) = +1$. Since the two limits do not equal, $\lim_{x \to 0} f(x)$ does not exist. This can be confirmed graphically.

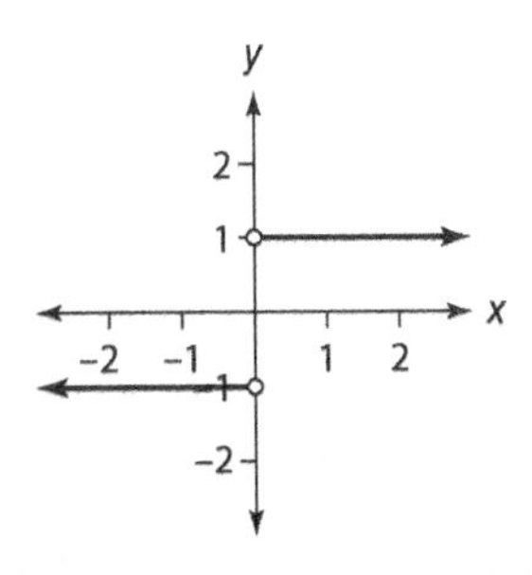

BRAIN TICKLERS Set # 6

Evaluate each of the following limits.

1. $\lim\limits_{x \to 4} \dfrac{1}{(x-4)^2}$
2. $\lim\limits_{x \to +\infty} e^{-2x}$
3. $\lim\limits_{x \to +\infty} \log_7 x$
4. $\lim\limits_{x \to -\infty} |x-5|$

(Answers are on page 34.)

Continuity

For a function, $f(x)$, to be continuous at a point where $x = a$, three conditions must be met.

1. The limit must exist for some real number L.

$$\lim_{x \to a} f(x) = L$$

2. The function value must exist for some real number L.

$$f(a) = L$$

3. The limit and the function value must be equal.

$$\lim_{x \to a} f(x) = f(a)$$

If any of the three criteria are not met, the function is discontinuous (not continuous) at that point.

The graph of $y = f(x)$ below displays two different types of discontinuities.

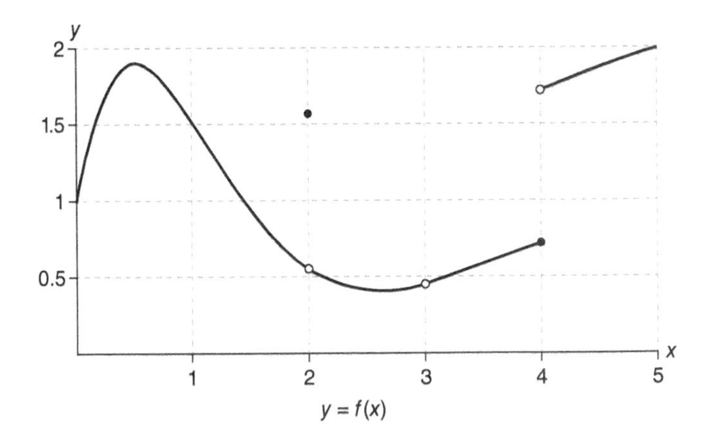

At $x = 2$ and at $x = 3$, the function has a *removable discontinuity*. At each point, the limit exists. However, the function value either does not exist, like at $x = 3$, or the function value does not equal the limit value, like at $x = 2$.

At $x = 4$, there is a *jump discontinuity*. The limit on the left as x approaches 4 is 0.75, and the limit on the right as x approaches 4 is 1.75. Since the limit does not exist, the function is not continuous at $x = 4$. It is described as a jump discontinuity since your hand "jumps" as it moves along the graph near $x = 4$.

The graph of $f(x) = \frac{1}{x^2}$ shows an *infinite discontinuity* at $x = 0$. However, the graph of $y = \sin\frac{1}{x}$ has an *oscillating discontinuity* at $x = 0$.

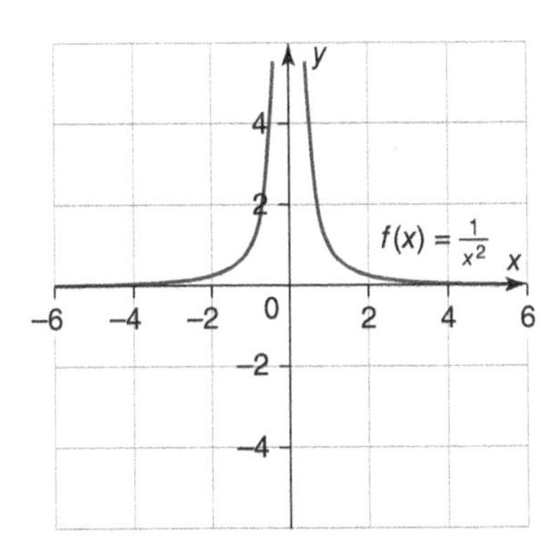

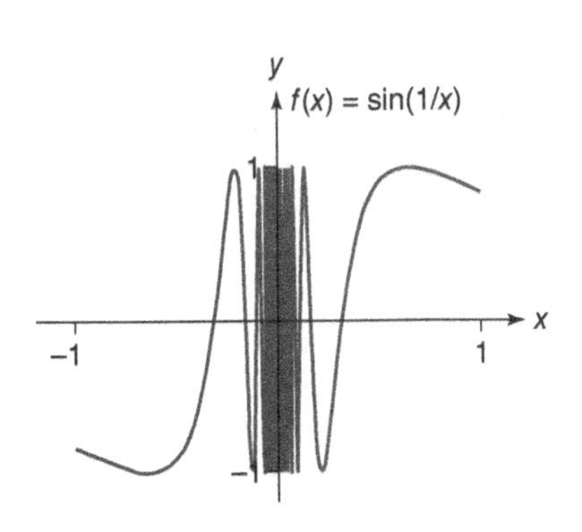

Example 20:

Is $f(x) = x^2 + 6x + 3$ continuous at $x = 1$?

Solution:

Determine whether the three conditions for a continuous function are met.

1. $\lim\limits_{x \to 1} x^2 + 6x + 3 = (1)^2 + 6(1) + 3 = 10$
2. $f(1) = (1)^2 + 6(1) + 3 = 10$
3. $\lim\limits_{x \to 1} x^2 + 6x + 3 = f(1)$

The function $f(x)$ is continuous at $x = 1$.

Since all polynomial functions have the same limit value as their function value, all polynomial functions are continuous for all real numbers. All polynomial functions are continuous functions.

Example 21:

Is $f(x) = \begin{cases} \dfrac{x^2 - 9}{x - 3}, & x \neq 3 \\ 6, & x = 3 \end{cases}$ continuous at $x = 3$?

Solution:

Determine whether the three conditions for a continuous function are met.

1. $\lim\limits_{x \to 3} f(x) = \lim\limits_{x \to 3} \dfrac{x^2 - 9}{x - 3} = \dfrac{(3)^2 - 9}{(3) - 3} = \dfrac{0}{0}$, which is indeterminate. Factoring and simplifying,
   $\lim\limits_{x \to 3} \dfrac{x^2 - 9}{x - 3} = \lim\limits_{x \to 3} \dfrac{(x + 3)(x - 3)}{(x - 3)} = \lim\limits_{x \to 3}(x + 3) = 6.$
2. $f(3) = 6$
3. $\lim\limits_{x \to 3} f(x) = f(3)$

The function $f(x)$ is continuous at $x = 3$.

1+2=3 MATH TALK!

Factoring out the common factor in Example 21 demonstrates a removable discontinuity since the common factor was "removed," allowing the limit to be evaluated successfully.

Example 22:

What value of k would make the function $f(x) = \begin{cases} \dfrac{x^2 - 25}{x - 5}, & x \neq 5 \\ k, & x = 5 \end{cases}$ continuous at $x = 5$?

Solution:

1. $\lim_{x \to 5} f(x) = \lim_{x \to 5} \dfrac{x^2 - 25}{x - 5} = \dfrac{0}{0}$, which is indeterminate. Factoring and simplifying,
   $\lim_{x \to 5} \dfrac{x^2 - 25}{x - 5} = \lim_{x \to 5} \dfrac{(x + 5)(x - 5)}{(x - 5)} = \lim_{x \to 5}(x - 5) = 10.$
2. $f(5) = k$
3. $\lim_{x \to 5}(x + 5) = f(5)$; therefore, $10 = k$.

If a function, $f(x)$, is continuous at each number in an open interval (a, b), then $f(x)$ is continuous on (a, b). If $f(x)$ is continuous on the open interval $(-\infty, \infty)$, then $f(x)$ is continuous everywhere. For example, polynomial functions and $f(x) = |x|$ are continuous everywhere.

For a function to be continuous at each number in a closed interval $[a, b]$, then $f(x)$ is continuous on the open interval (a, b) and its value at each endpoint is equal to the appropriate one-sided limit at that endpoint.

1+2=3 MATH TALK!

For $f(x)$ to be continuous over $[a, b]$, the following conditions must be met:

1. $f(x)$ is continuous over (a, b)
2. $\lim_{x \to a^+} f(x) = f(a)$
3. $\lim_{x \to b^-} f(x) = f(b)$

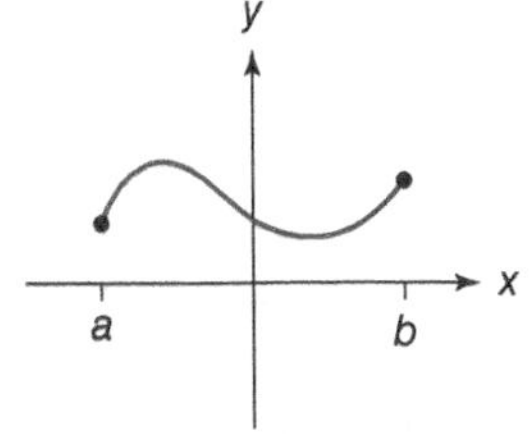

Example 23:

Describe the continuity of the function $f(x) = \sqrt{9 - x^2}$.

Solution:

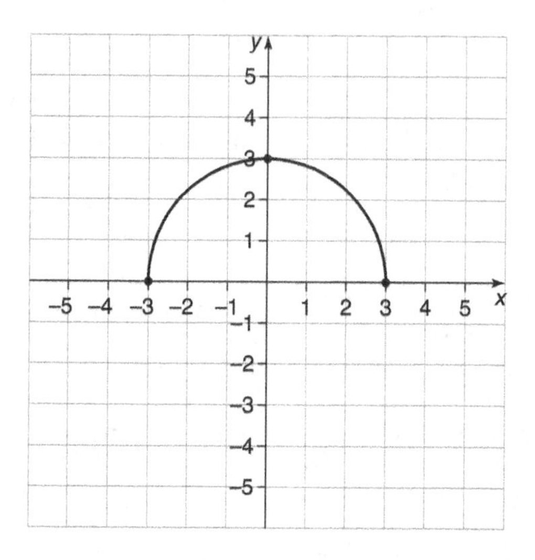

Using the graph of $f(x)$, the following can be seen.

1. $f(x)$ is continuous over $(-3, 3)$.
2. $\lim\limits_{x \to -3^+} f(x) = 0 = f(-3)$
3. $\lim\limits_{x \to 3^-} f(x) = 0 = f(3)$

$f(x) = \sqrt{9 - x^2}$ is continuous over $[-3, 3]$.

Intermediate Value Theorem

The Intermediate Value Theorem states that if $f(x)$ is continuous on a closed interval $[a, b]$ and if C is any number between $f(a)$ and $f(b)$, inclusive, then there is at least one number x in the interval $[a, b]$ such that $f(x) = C$.

1+2=3 MATH TALK!

The Intermediate Value Theorem, also referred to as the **IVT**, is one of the existence theorems in calculus. What is interesting about the IVT is that it uses function values in the range to prove the existence of a corresponding x-value in the domain. Think of the IVT as a diet, where your starting weight is 170 pounds and you achieve your goal weight of 140 pounds on the 75th day. The IVT guarantees that you would weigh 160 pounds on some day between the first and 75th day.

The Intermediate Value Theorem can be particularly useful if the goal is to find zeros (the x-values that make a polynomial function equal to 0) of polynomial functions. The IVT proves the existence of zeros in an interval.

Example 24:

Show that the function $f(x) = x^2 + x - 1$ has at least one zero in the interval $[-1, 5]$.

Solution:

Since $f(x) = x^2 + x - 1$ is a polynomial function, it is continuous everywhere, including the interval $[-1, 5]$. Evaluating the function at the endpoints of the interval gives $f(-1) = -1$ and $f(5) = 29$. Since 0 is between -1 and 29, by the IVT there exists at least one x in $[-1, 5]$ where $f(x) = 0$.

BRAIN TICKLERS Set # 7

1. Find the value of k that would make $f(x) = \begin{cases} \dfrac{x^2 - 9}{x - 3}, & x \neq 3 \\ 2k + 1, & x = 3 \end{cases}$ continuous at $x = 3$.
2. Describe why $g(x) = \begin{cases} \dfrac{3}{x^2}, & x \neq 0 \\ 2, & x = 0 \end{cases}$ is not continuous at $x = 0$.
3. Describe why $f(x) = \tan x$ is not continuous on $\left[0, \dfrac{\pi}{2}\right]$.
4. The function $f(x)$ is continuous on the closed interval $[1, 10]$. If $f(1) = 1$, $f(5) = c$, and $f(10) = 2$, then the equation $f(x) = \dfrac{1}{2}$ must have at least two solutions in the interval $[1, 10]$ if c equals what value?

(Answers are on page 34.)

BRAIN TICKLERS—THE ANSWERS

Set # 1, page 8

1. 1
2. $+\infty$ or does not exist
3. –1
4. 0

Set # 2, page 12

1. –2
2. 3
3. 112
4. 0

Set # 3, page 16

1. 2
2. $\frac{3}{4}$
3. 8
4. 4

Set # 4, page 20

1. $\frac{1}{4}$
2. 2
3. 0
4. 3

Set # 5, page 25

1. $-\infty$
2. $-\infty$
3. 0
4. Does not exist or $+\infty$

Set # 6, page 28

1. Does not exist or $+\infty$
2. 0
3. $+\infty$
4. $+\infty$

Set # 7, page 33

1. $k = \frac{5}{2}$
2. $g(x)$ is not continuous at $x = 0$ because $\lim_{x\to 0} g(x)$ does not exist.
3. $f(x) = \tan x$ is not continuous on $\left[0, \frac{\pi}{2}\right]$ because $\lim_{x\to \frac{\pi}{2}^-} \tan x$ does not exist.
4. If c is any real number less than $\frac{1}{2}$, then $f(x) = \frac{1}{2}$ must have at least two solutions in the interval.

Chapter 2

Differentiation

Differential calculus has two focuses. The first is to find the slope of a line tangent to a curve at a point. The second is to find the rate of change in one variable with respect to another.

Average Rate of Change

The average rate of change is the change in the value of a quantity divided by the change in the value of another quantity. Common examples of average rate of change include finding the average speed $= \dfrac{\text{change in distance}}{\text{change in time}}$ and the slope of a line segment $= \dfrac{\Delta y}{\Delta x}$.

1+2=3 MATH TALK!

To find any type of average rate of change, you must have two given points or two sets of inputs with corresponding outputs in order to take the difference.

Finding the average rate of change between an initial point and a final point is equivalent to finding the slope of the segment (or secant) between those points.

Example 1:

A car travels between two cities according to the following the distance-time graph.

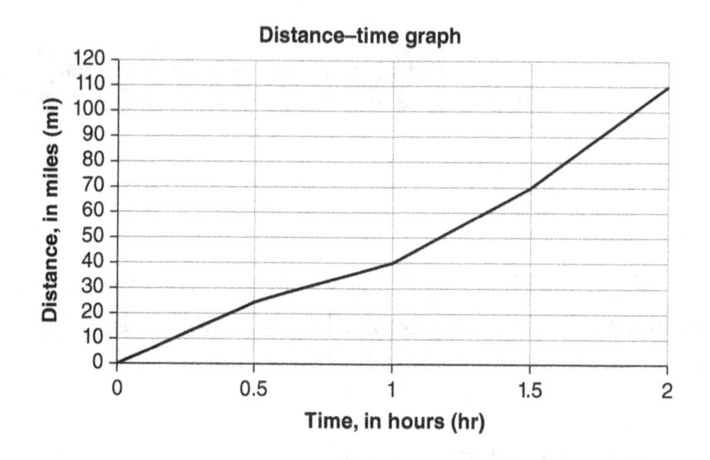

1. What is the car's average speed for the first hour?
2. What is the car's average speed for the entire trip?
3. How does the average speed in question 1 compare to the average speed in question 2?

Solution:

1. Initially, the car has traveled 0 miles in 0 hours.

   After 1 hour, the car has traveled 40 miles.

   The average speed $= \frac{40 - 0}{1 - 0} = 40$ mph.

2. Initially, the car has traveled 0 miles in 0 hours.

   By the end of the trip, the car has traveled 110 miles in 2 hours.

   The average speed $= \frac{110 - 0}{2 - 0} = 55$ mph.

3. Draw the two secants that represent the solutions in questions 1 and 2, as shown on the following page. The slope of the secant for question 2 is steeper than the slope of the secant for question 1. In other words, over the course of the entire trip, more distance was gained after the initial hour.

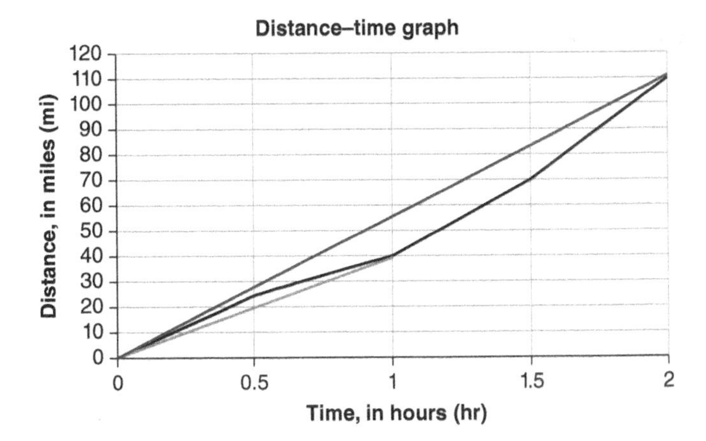

The average speed for the entire trip, as described in Example 1, does not give all the details for the speeds along the way. The average speed is just that, an average, or an overall idea of the rate of change between a starting point and an ending point.

Finding the speed at a specific instant in time is something entirely different. Graphically, at one point, such as $t = 1$ hour, the secant line from Example 1 would change to a tangent line, as shown in the figure below.

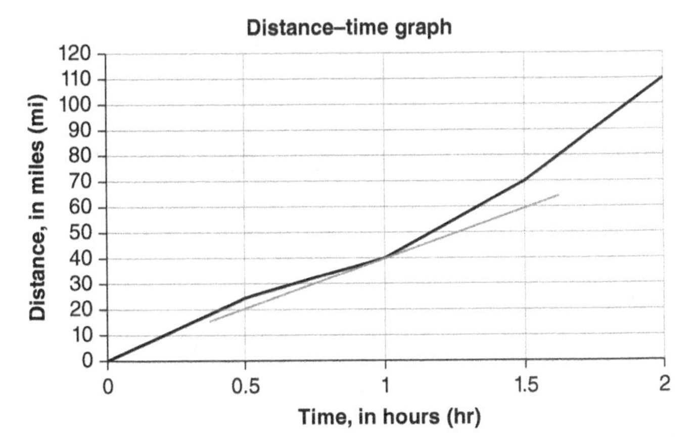

Finding the rate of change at this time is an issue since the previous slope formula needs two points and only one point is known for the tangent line.

Slope of a Tangent Line

A line tangent to a curve is a straight line that intersects the curve at a particular point and has the same slope as the curve at that point. Example 2 will investigate how the slopes of secant lines approach the value of the slope of a tangent line.

Example 2:

Given the function $f(x) = x^2$, find the slope of the tangent line to $f(x)$ at $x = 4$.

Solution:

At $x = 4$, the coordinates of the point on $f(x)$ are (4, 16). Since using the slope formula $\frac{\Delta y}{\Delta x}$ is impossible at only one point, the slopes of secant lines as they get closer to the point (4, 16) will be evaluated to see if there is a pattern.

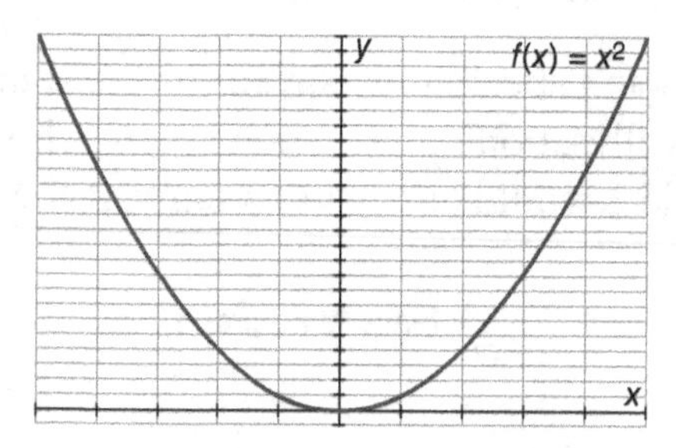

Secant Endpoints	Slope of the Secant
(0, 0) and (4, 16)	$\frac{16-0}{4-0} = 4$
(1, 1) and (4, 16)	$\frac{16-1}{4-1} = 5$
(2, 4) and (4, 16)	$\frac{16-4}{4-2} = 6$
(3, 9) and (4, 16)	$\frac{16-9}{4-3} = 7$
(3.5, 12.25) and (4, 16)	$\frac{16-12.25}{4-3.5} = 7.5$
(4, 16) and (4.5, 20.25)	$\frac{20.25-16}{4.5-4} = 8.5$
(4, 16) and (5, 25)	$\frac{25-16}{5-4} = 9$

When starting on the left at the point $(0, 0)$ and moving right toward the point $(4, 16)$, the slopes of the secant lines approach 8. When starting on the right at the point $(5, 25)$ and moving left toward $(4, 16)$, the slopes of the secant lines also approach 8. This is similar to how limits are evaluated, and it appears that the slope of the line tangent to $f(x) = x^2$ at $x = 4$ is 8.

1+2=3 MATH TALK!

Drawing in each of the secants from the table above on the graph of x^2, as the x-values approach 4, would show the secants resembling the tangent line at $x = 4$.

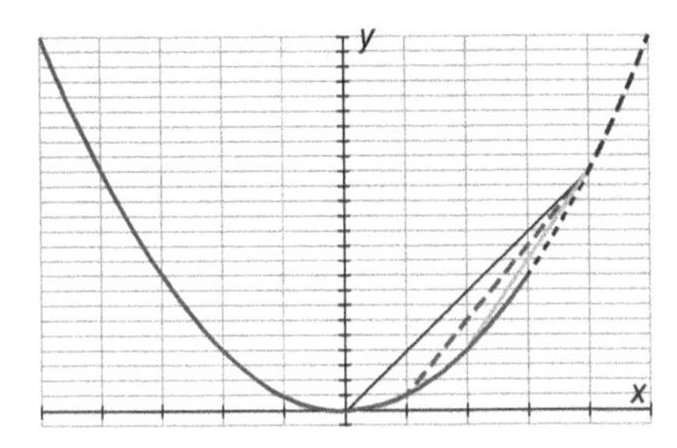

In general, the formula to find the slope of a tangent line to a function $f(x)$ at $x = a$ is

$$\text{Slope of a tangent line} = \lim_{x \to a} \frac{f(x) - f(a)}{x - a}$$

The expression $\frac{f(x) - f(a)}{x - a}$ represents the slope of a secant line. To make it the slope of a tangent line, the limit is introduced to have the general x-value approach a but not equal to a; otherwise, the expression would be indeterminate.

The slope of the tangent line is also referred to as the instantaneous rate of change of a function at a particular point.

The limit can also be notated as the following:

$$f'(a) = \lim_{x \to a} \frac{f(x) - f(a)}{x - a}$$

The notation $f'(a)$ is read "f prime of a" and represents the derivative of $f(x)$ at $x = a$.

1+2=3 MATH TALK!

The slope of a tangent line has other meanings. The slope of a tangent line is the same value as the instantaneous rate of change of a function at that point and the derivative of a function at that point. To find all three meanings, use the same limit expression $f'(a) = \lim_{x \to a} \frac{f(x) - f(a)}{x - a}$.

Example 3:

Given the function $f(x) = x^2$, find $f'(4)$.

Solution:

$$f'(4) = \lim_{x \to 4} \frac{f(x) - f(4)}{x - 4} = \lim_{x \to 4} \frac{x^2 - 16}{x - 4} = \frac{0}{0}$$

This is indeterminate. Factoring the expression and simplifying,

$$\lim_{x \to 4} \frac{x^2 - 16}{x - 4} = \lim_{x \to 4} \frac{(x + 4)(x - 4)}{x - 4} = \lim_{x \to 4} x + 4 = 8$$

Therefore, $f'(4) = 8$, which also means that the slope of the tangent line of $f(x) = x^2$ at $x = 4$ is also 8, and the instantaneous rate of change of the function at $x = 4$ is 8.

Example 4:

Find the slope of the tangent line to $f(x) = 2x^2 + 5$ at the point where $x = 3$.

Solution:

The slope of the tangent line at a point is found using $f'(a) = \lim_{x \to a} \frac{f(x) - f(a)}{x - a}$, where $f(x) = 2x^2 + 5$ and $a = 3$:

$$f'(3) = \lim_{x \to 3} \frac{f(x) - f(3)}{x - 3} = \lim_{x \to 3} \frac{2x^2 + 5 - (2(3^2) + 5)}{x - 3}$$

$$= \lim_{x \to 3} \frac{2x^2 + 5 - 23}{x - 3} = \lim_{x \to 3} \frac{2x^2 - 18}{x - 3} = \frac{0}{0}$$

Factoring and simplifying give the slope:

$$\lim_{x\to 3}\frac{2x^2-18}{x-3}=\lim_{x\to 3}\frac{2(x+3)(x-3)}{x-3}=\lim_{x\to 3}(2(x+3))=12$$

This can be verified graphically. Graphing the function $f(x)=2x^2+5$, plotting the point where $x=3$, and sketching the tangent line at this point show the slope is positive and can be 12.

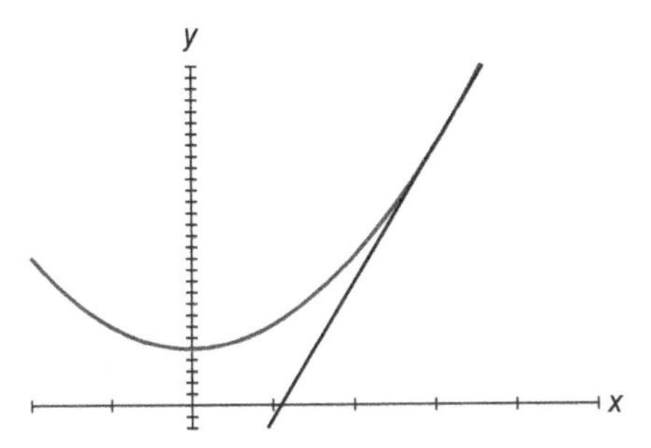

Example 5:

Find the slope of the tangent line to the curve whose equation is $f(x)=5$ at the point where $x=-2$.

Solution:

The slope of the tangent line at a point is found using $f'(a)=\lim_{x\to a}\frac{f(x)-f(a)}{x-a}$, where $f(x)=5$ and $a=-2$:

$$f'(-2)=\lim_{x\to -2}\frac{f(x)-f(-2)}{x-(-2)}=\lim_{x\to -2}\frac{5-5}{x+2}$$

$$=\lim_{x\to -2}\frac{0}{x+2}=\lim_{x\to -2}0=0$$

This can be verified graphically. Graphing the function $f(x)=5$, plotting the point where $x=-2$, and graphing the tangent line at this point show it is a horizontal line. The slope of a horizontal line is equal to 0, which agrees with the limit definition answer.

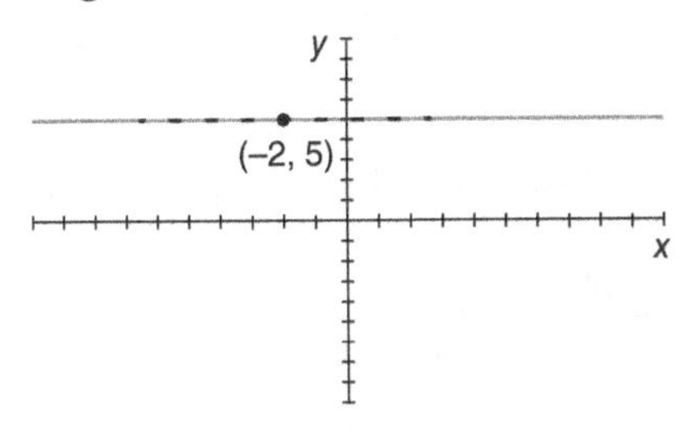

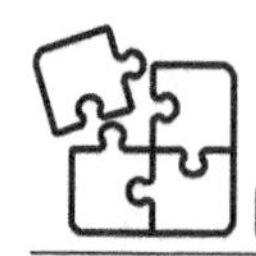

BRAIN TICKLERS Set # 8

1. For $f(x) = 3x^2$, find the average rate of change of the function over the interval [0, 2].
2. Using the table of values below, approximate $f'(3.5)$ using the average rate of change.

x	0	1	2	3	4	5	6
$f(x)$	0	5.3	8.8	11.2	12.8	13.8	14.5

3. Find the slope of the tangent line of the function $f(x) = 4x + 7$ at $x = 3$.
4. Given $f(x) = x^2$, find the instantaneous rate of change of the function when $x = 3$.

(Answers are on page 52.)

Another way to calculate the derivative at a point, $x = a$, is to use the following limit definition:

$$f'(a) = \lim_{h \to 0} \frac{f(a+h) - f(a)}{h}$$

Example 6:

Using the formula $f'(a) = \lim_{h \to 0} \frac{f(a+h) - f(a)}{h}$, find the instantaneous rate of change in $f(x) = x^2 + 2x$ at the point where $x = 4$.

Solution:

$$f'(4) = \lim_{h \to 0} \frac{f(4+h) - f(4)}{h}$$

$$= \lim_{h \to 0} \frac{(4+h)^2 + 2(4+h) - ((4)^2 + 2(4))}{h}$$

$$= \lim_{h \to 0} \frac{16 + 8h + h^2 + 8 + 2h - 24}{h} = \lim_{h \to 0} \frac{h^2 + 10h}{h} = \frac{0}{0}$$

Factoring and simplifying give the instantaneous rate of change:

$$\lim_{h \to 0} \frac{h^2 + 10h}{h} = \lim_{h \to 0} \frac{h(h+10)}{h} = \lim_{h \to 0}(h + 10) = 10$$

1+2=3 **MATH TALK!**

Unless otherwise specified, either limit definition for $f'(a)$ can be used if asked to find the slope of a tangent line at $x = a$, the instantaneous rate of change at $x = a$, or the derivative of a function at $x = a$.

Definition of a Derivative Function

Instead of finding the derivative value at one point for a function, there is a way to find a derivative function. Once a derivative function is found, it can be used to find the derivative at any x-value.

The definition of a *derivative function* is

$$f'(x) = \lim_{h \to 0} \frac{f(x+h) - f(x)}{h}$$

It is similar to our second definition except a general x-variable is substituted in for the a-value that represents a number. This is what makes it a derivative function of x.

Example 7:

Use the definition of $f'(x)$ to find the slope function for $f(x) = x^2$.

Solution:

$$f'(x) = \lim_{h \to 0} \frac{f(x+h) - f(x)}{h} = \lim_{h \to 0} \frac{(x+h)^2 - x^2}{h}$$

$$= \lim_{h \to 0} \frac{x^2 + 2xh + h^2 - x^2}{h} = \lim_{h \to 0} \frac{2xh + h^2}{h} = \frac{0}{0}$$

After factoring and simplifying:

$$\lim_{h \to 0} \frac{2xh + h^2}{h} = \lim_{h \to 0} \frac{h(2x+h)}{h} = \lim_{h \to 0}(2x + h) = 2x$$

Discovering that $f'(x) = 2x$ when $f(x) = x^2$ is very useful. For any x-value, the slope of a tangent can be calculated by substituting that value into $f'(x)$. This can be demonstrated using the graph of $f(x) = x^2$.

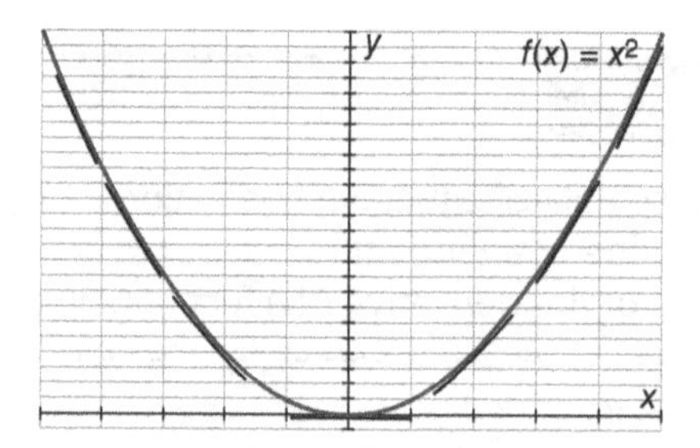

When $x < 0$, $f'(x) < 0$. Graphically, the tangent lines all have a negative slope, matching the derivative values.

When $x = 0$, $f'(x) = 0$. Graphically, the tangent line has a zero slope, matching the derivative value.

When $x > 0$, $f'(x) > 0$. Graphically, the tangent lines all have a positive slope, matching the derivative values.

CAUTION—Major Mistake Territory!

There is a difference between $f'(a)$ and $f'(x)$. When finding $f'(a)$, the solution will be a numerical value that represents the slope of the tangent line at that specific x-value. When finding $f'(x)$, the solution will be a function of x that represents the derivative function. After finding $f'(x)$, multiple x-values can be substituted into this function to find their corresponding slopes.

All three limit definitions can be used when trying to find the slope or instantaneous rate of change at a specific x-value. However, if choosing to find $f'(x)$, you must remember to substitute the x-value into the $f'(x)$ function.

Example 8:

Find the derivative function, $f'(x)$, for $f(x) = \sqrt{x}$.

Solution:

$$f'(x) = \lim_{h \to 0} \frac{f(x+h) - f(x)}{h} = \lim_{h \to 0} \frac{\sqrt{x+h} - \sqrt{x}}{h} = \frac{0}{0}$$

This is indeterminate. Multiplying by the conjugate of the numerator and simplifying give the following:

$$\lim_{h\to 0}\frac{\sqrt{x+h}-\sqrt{x}}{h}\bullet\frac{\sqrt{x+h}+\sqrt{x}}{\sqrt{x+h}+\sqrt{x}}=\lim_{h\to 0}\frac{x+h-x}{h\left(\sqrt{x+h}+\sqrt{x}\right)}$$

$$=\lim_{h\to 0}\frac{1}{\sqrt{x+h}+\sqrt{x}}=\frac{1}{\sqrt{x}+\sqrt{x}}=\frac{1}{2\sqrt{x}}$$

REMINDER

The derivative of a function at a point has two other meanings:

1. The derivative is also the instantaneous rate of change of the function at that value.
2. The derivative is also the slope of the tangent line to the graph of the function at that value.

There are three ways to find the derivative of a function at a point:

1. $f'(a)=\lim_{x\to a}\frac{f(x)-f(a)}{x-a}$
2. $f'(a)=\lim_{h\to 0}\frac{f(a+h)-f(a)}{h}$
3. $f'(x)=\lim_{h\to 0}\frac{f(x+h)-f(x)}{h}$. However, this is the derivative function. To find a specific derivative at a point, the x-value must be substituted into the derivative function.

BRAIN TICKLERS Set # 9

1. Find $f'(5)$ if $f(x)=x^2+2x$.
2. Using the definition of the derivative, find the derivative of $f(x)=x^2+1$ when $x=1$.
3. Find the slope of the line tangent to the curve whose function is $g(x)=4x^3-x^2+\frac{x}{2}+7$ at $x=-2$.
4. If $f(x)=x^2-5x+1$, find $f'(x)$.

(Answers are on page 52.)

Differentiability

Finding the derivative function allows the slope of a tangent line to a curve to be found at an x-value. The question is, will the derivative exist at all x-values?

In Example 8, the derivative function for $f(x) = \sqrt{x}$ was $f'(x) = \frac{1}{2\sqrt{x}}$. At $x = 0$, $f'(0) = \frac{1}{2\sqrt{0}} = \frac{1}{0}$, which is undefined. In other words, $f'(0)$ does not exist. That is what is meant by differentiability. For a function to be differentiable, the derivative of a function at that point exists.

1+2=3 MATH TALK!

When asked if a function is differentiable at a value, it means that the derivative exists at that value. If a function is not differentiable at a point, graphically a tangent line is either a vertical line or is unable to be graphed at that value.

Example 9:

Given $f(x) = |x|$, prove algebraically that $f(x)$ is not differentiable at $x = 0$.

Solution:

Rewrite $f(x) = |x|$ as a piecewise function:

$$f(x) = \begin{cases} -x, \ x < 0 \\ 0, \ x = 0 \\ x, \ x > 0 \end{cases}$$

Evaluate $f'(0)$:

$$f'(0) = \lim_{x \to 0} \frac{f(x) - f(0)}{x - 0}$$

Since $x = 0$ is on the boundary, the left-hand and right-hand limits need to be evaluated.

Left-Hand Limit	Right-Hand Limit
$\lim\limits_{x \to 0^-} \frac{f(x) - f(0)}{x - 0} = \lim\limits_{x \to 0^-} \frac{-x - 0}{x - 0}$	$\lim\limits_{x \to 0^+} \frac{f(x) - f(0)}{x - 0} = \lim\limits_{x \to 0^+} \frac{x - 0}{x - 0}$
$= \lim\limits_{x \to 0^-} \frac{-x}{x} = \lim\limits_{x \to 0^-} -1 = -1$	$= \lim\limits_{x \to 0^+} \frac{x}{x} = \lim\limits_{x \to 0^+} 1 = 1$

Since the left-hand and right-hand limits are not equal, the limit does not exist. Therefore, $f'(0)$ does not exist if $f(x) = |x|$.

The graph of $f(x) = |x|$ shows a corner at $x = 0$. Since the function is not differentiable at this value, it is impossible to draw a tangent line at $x = 0$.

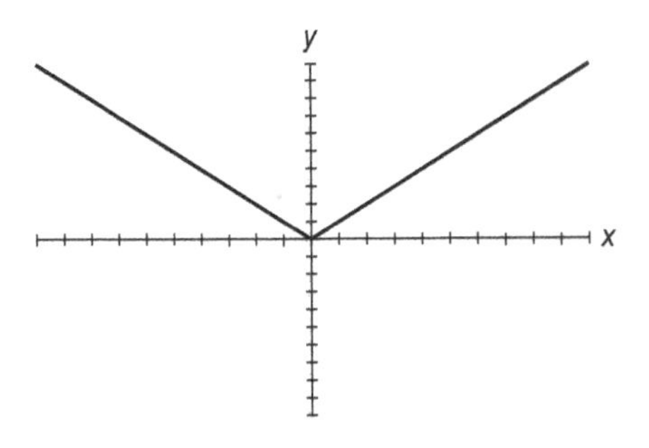

Example 10:

Given $f(x) = \begin{cases} x^2 - 5, & x \leq 1 \\ x - 5, & x > 1 \end{cases}$, is the function differentiable at $x = 1$?

Solution:

Since $x = 1$ is at the boundary of the piecewise function, the left-hand and right-hand limits need to be evaluated.

Left-Hand Limit	Right-Hand Limit
$\lim_{x \to 1^-} \frac{f(x) - f(1)}{x - 1} = \lim_{x \to 1^-} \frac{x^2 - 5 - ((1)^2 - 5)}{x - 1}$	$\lim_{x \to 1^+} \frac{f(x) - f(1)}{x - 1}$
$= \lim_{x \to 1^-} \frac{x^2 - 5 - (-4)}{x - 1}$	$= \lim_{x \to 1^+} \frac{x - 5 - ((1)^2 - 5)}{x - 1}$
$= \lim_{x \to 1^-} \frac{x^2 - 1}{x - 1}$	$= \lim_{x \to 1^+} \frac{x - 5 - (-4)}{x - 1}$
$= \lim_{x \to 1^-} \frac{(x + 1)(x - 1)}{x - 1}$	$= \lim_{x \to 1^+} \frac{x - 1}{x - 1}$
$= \lim_{x \to 1^-} (x + 1)$	$= \lim_{x \to 1^+} 1$
$= 2$	$= 1$

Since the left-hand and right-hand limits are not equal, the limit does not exist. Therefore, $f'(1)$ does not exist for the given piecewise function.

The graph of $f(x) = \begin{cases} x^2 - 5, & x \leq 1 \\ x - 5, & x > 1 \end{cases}$ shows a cusp at $x = 1$. Since the function is not differentiable at this value, it is impossible to draw a tangent line at $x = 0$.

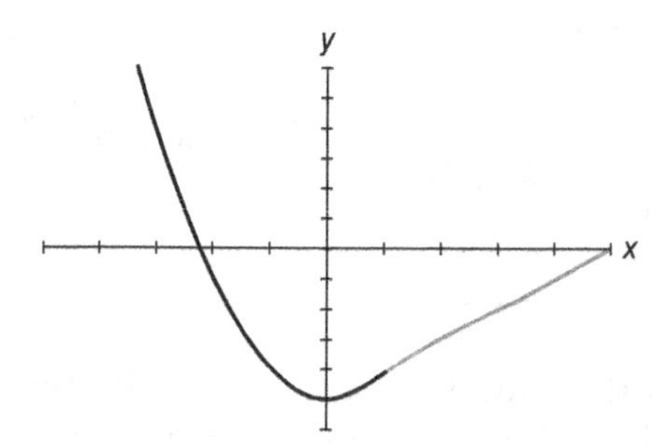

Example 11:

Given $f(x) = \begin{cases} \frac{x^2 - 4}{x - 2}, & x \neq 2 \\ 12, & x = 2 \end{cases}$, is this function differentiable at $x = 2$?

Solution:

$$f'(2) = \lim_{x\to 2}\frac{f(x) - f(2)}{x - 2} = \lim_{x\to 2}\frac{\frac{x^2 - 4}{x - 2} - 12}{x - 2}$$

$$= \lim_{x\to 2}\frac{\frac{(x + 2)(x - 2)}{x - 2} - 12}{x - 2} = \lim_{x\to 2}\frac{x + 2 - 12}{x - 2}$$

$$= \frac{2 + 2 - 12}{2 - 2} = \frac{-8}{0}$$

This value is undefined. Therefore, $f'(2)$ does not exist for this function.

Look at the graph of $f(x) = \begin{cases} \frac{x^2 - 4}{x - 2}, x \neq 2 \\ 12, x = 2 \end{cases}$. At $x = 2$, there is a hole (removable discontinuity). Since the function is not differentiable at this value, it is impossible to draw a tangent line at $x = 2$.

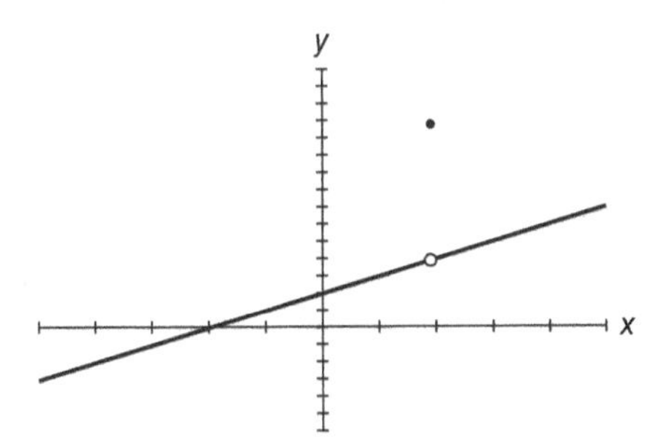

Example 12:

If $f(x) = x^{1/3}$, is $f(x)$ differentiable at $x = 0$?

Solution:

$$f'(0) = \lim_{x\to 0}\frac{f(x) - f(0)}{x - 0} = \lim_{x\to 0}\frac{x^{1/3} - 0}{x - 0} = \lim_{x\to 0}\frac{x^{1/3}}{x} = \lim_{x\to 0}\frac{1}{x^{2/3}} = \frac{1}{0}$$

This value is undefined. Therefore, $f'(0)$ does not exist for this function.

Look at the graph of $f(x) = x^{1/3}$. At $x = 0$, there is a vertical tangent. The slopes of vertical lines are undefined.

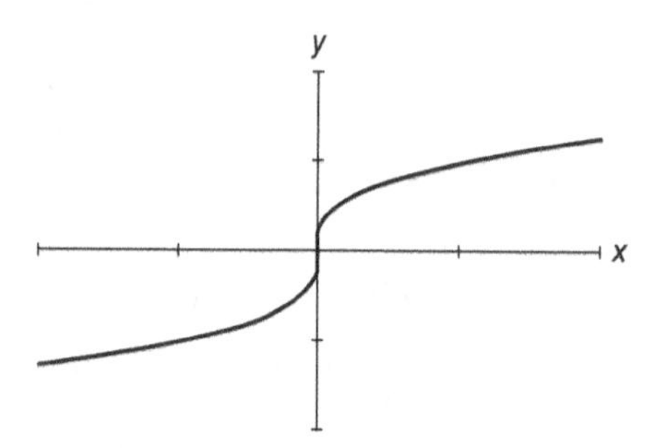

Example 13:

If $f(x) = x^{4/3}$, is $f(x)$ differentiable at $x = 0$?

Solution:

$$f'(0) = \lim_{x\to 0} \frac{f(x) - f(0)}{x - 0} = \lim_{x\to 0} \frac{x^{4/3} - 0}{x - 0} = \lim_{x\to 0} \frac{x^{4/3}}{x} = \lim_{x\to 0} x^{1/3} = 0$$

The limit exists, and therefore the function is differentiable at $x = 0$.

Look at the graph of $f(x) = x^{4/3}$. At $x = 0$, the tangent is horizontal. The slopes of horizontal lines are equal to zero.

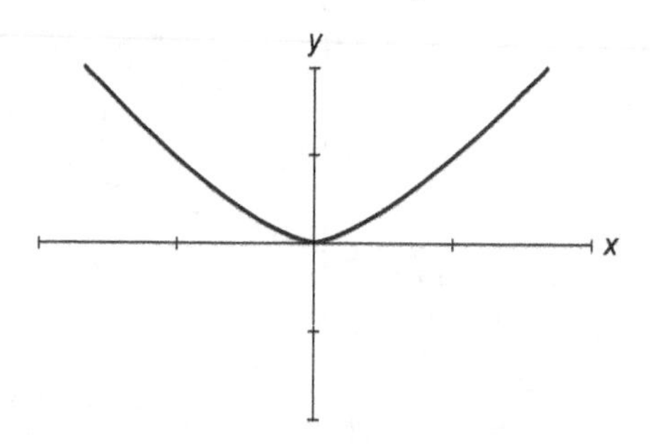

1+2=3 MATH TALK!

For a function to be differentiable, the derivative must exist at that point. Graphically, there are ways to tell if a function is not differentiable. On a graph, a function is not differentiable at a point if at that point there is a:

1. corner
2. cusp
3. hole
4. vertical asymptote
5. vertical tangent

Continuity and Differentiability

Chapter One explains how to discover if a function is continuous at a point. This feels similar to finding whether a function is differentiable at a point. Two questions naturally arise: "Does continuity imply differentiability? Does differentiability imply continuity?"

The answer to the first question is no. Continuity does *not* imply differentiability. For example, the function $f(x) = |x|$ is continuous for all x; however, it is not differentiable at $x = 0$.

The answer to the second question, however, is yes. Differentiability *does* imply continuity. If a function is differentiable at a point, over an interval, or for all values of x, it is also continuous at those values.

BRAIN TICKLERS Set # 10

1. Find $f'(x)$ if $f(x) = 2x^2 + 1$.
2. Given $f(x) = (x - 2)^{2/3}$, state if the function is differentiable at $x = 2$.
3. Using the following graph of $f(x)$, state all values of x for which the function is not differentiable in the interval (0, 5).

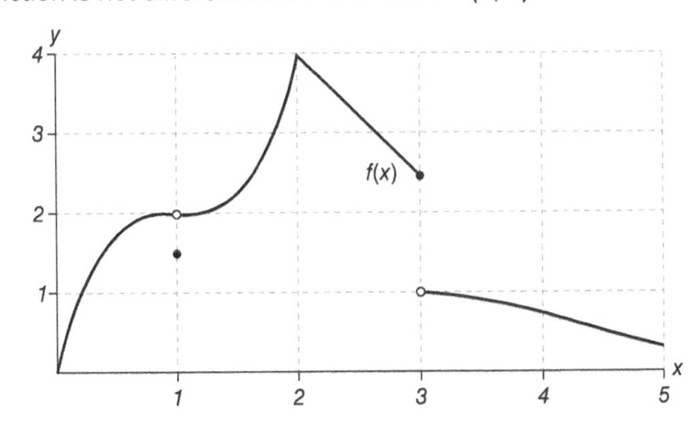

4. Given $f(x) = \begin{cases} x^2 + 1, x \leq 1 \\ 2x, x > 1 \end{cases}$, state if the function is continuous and/or differentiable at $x = 1$.

(Answers are on page 52.)

BRAIN TICKLERS—THE ANSWERS

Set # 8, page 42

1. 6
2. 1.6
3. 4
4. 6

Set # 9, page 45

1. 12
2. 2
3. 52.5
4. $2x - 5$

Set # 10, page 51

1. $f'(x) = 4x$
2. The function $f(x) = (x - 2)^{2/3}$ is not differentiable at $x = 2$ because the derivative does not exist.
3. The function is not differentiable at $x = 1, 2,$ and 3.
4. The function $f(x) = \begin{cases} x^2 + 1, \; x \le 1 \\ 2x, x > 1 \end{cases}$ is differentiable at $x = 1$ and therefore is also continuous at $x = 1$.

Chapter 3

Derivatives

Now that you understand that the *derivative of a function* represents both the slope of a tangent line and the instantaneous rate of change, this chapter will focus on using shortcut rules to find the derivatives of different functions. In Chapter Two, the derivative of a function was found by using the *limit definition*. Although this will always work, it can be very challenging and time-consuming as functions become more complicated.

Notation

There are a few different ways to notate the derivative of a function, as shown in the table below.

Notation	Read as
$y'(x)$ or y'	y prime of x or y prime
$\frac{dy}{dx}$	derivative of y with respect to x or "dee y, dee x"
$f'(x)$ or f'	f prime of x or f prime
$\frac{d}{dx}(f(x))$ or $\frac{d}{dx}(f)$	derivative of f of x with respect to x or derivative of f with respect to x

Not all functions will be written using the variables x and y. For example, if you are working with a function that represents velocity over time, it might be notated as $v(t)$ instead of $f(x)$. The derivative notation would change as well and would be written as $v'(t)$ or $\frac{dv}{dt}$.

The prime notation is more commonly used. However, it is better to use $\frac{dy}{dx}$ since it reinforces the fact that a derivative is a slope. Think of $\frac{dy}{dx}$ like $\frac{\Delta y}{\Delta x}$.

Using $\frac{dy}{dx}$ also specifies the independent variable the derivative is taken with respect to, in this case x. If a function simply says y', it is hard to know if y is a function of x as in $y(x)$ or of some other variable like t as in $y(t)$.

The notation shown in the table above represents the first derivative. If possible, you can further differentiate. The second derivative is found by differentiating the first derivative; the third derivative is found by differentiating the second derivative; and so on. The meaning of these higher-order derivatives will be discovered in future chapters.

1+2=3 MATH TALK!

Differentiation is the process of finding a derivative.

The order of a derivative is a positive integer that specifies how many times the function has been differentiated.

The second derivative can be notated as shown in the following table.

Notation	Read as
$y''(x)$ or y''	y double prime of x or y double prime
$\frac{d^2y}{dx^2}$	second derivative of y with respect to x
$f''(x)$ or f''	f double prime of x or f double prime

The third derivative is notated as y''', f''', or $\frac{d^3y}{dx^3}$. For a fourth derivative and higher-order derivatives, the notation changes. The fourth derivative is written as either $f^{(4)}(x)$ or $\frac{d^4y}{dx^4}$. In general, derivatives can be notated as $f^{(n)}(x)$ or $\frac{d^ny}{dx^n}$, where n is the order of the derivative.

CAUTION—Major Mistake Territory!

Do not confuse $y^{(n)}(x)$, which notates the nth derivative of y, with the nth power of y, which is $y^n(x)$.

Properties of Derivatives

Three derivative properties can be applied to the derivative shortcut rules.

Property 1: The derivative of a constant is equal to zero.

If a is a constant, then $\frac{d}{dx}(a) = 0$.

Property 2: The derivative of the sum or difference of two or more differentiable functions is equal to the sum or difference of their derivatives.

If u and v are differentiable functions of x, then $\frac{d}{dx}(u \pm v) = \frac{du}{dx} \pm \frac{dv}{dx}$.

Property 3: A constant can be factored out of a function and multiplied by the derivative of the function.

If a is a constant and u is a differentiable function of x, then $\frac{d(au)}{dx} = a \bullet \frac{du}{dx}$.

CAUTION—Major Mistake Territory!

Taking the derivative of a product or of a quotient of functions does not follow the sum and difference property.

$$\frac{d}{dx}(u \bullet v) \neq \frac{du}{dx} \bullet \frac{dv}{dx}$$

$$\frac{d}{dx}\left(\frac{u}{v}\right) \neq \frac{\frac{du}{dx}}{\frac{dv}{dx}}$$

They have their own special rules, which can be found on pages 59–62.

Power Rule

The first shortcut rule can be applied to polynomials or monomials raised to numerical exponents, such as $x^2 + x + 3, x^{1/2}$, or x^{-7}.

This shortcut, known as the *Power Rule*, can be generalized as:

If n is a constant and $x \neq 0$, then $\frac{d}{dx}(x^n) = n \bullet x^{n-1}$.

1+2=3 MATH TALK!

When applying the Power Rule, think of it as bringing the exponent down, multiplying it by the base, keeping the base the same, and then subtracting the exponent by 1.

If a constant is already in front of the base, multiply that constant by the exponent.

Finding derivatives is *painless*. You can use three steps to find the derivative of a function.

Step 1: Decide which derivative rule to use.

Step 2: Apply the rule.

Step 3: Simplify.

When applying Step 1, consider each term in the function and all operations. Use the derivative properties where appropriate and necessary.

Example 1:

Find $\frac{dy}{dx}$ if $y = -10x^5$.

Solution:

Step 1: Decide which derivative rule to use.

Use Property 3 since –10 is a constant that can be factored out and multiplied by the derivative of x^5. Since x^5 is a monomial with an exponent, apply the Power Rule.

Step 2: Apply the Power Rule:

$$\frac{dy}{dx} = -10 \bullet 5 \bullet x^{5-1}$$

Step 3: Simplify:

$$\frac{dy}{dx} = -50x^4$$

CAUTION—Major Mistake Territory!

If an expression involves radicals or fractions, do not stress! Rewrite the terms using exponents so that the Power Rule can be applied.

For example: $\sqrt{x} = x^{1/2}$ and $\frac{1}{x^7} = x^{-7}$.

Example 2:

Find $f'(x)$ if $f(x) = \sqrt{x}$.

Solution:

Step 1: Decide which derivative rule to use.

Rewrite the radical expression using exponents:

$$f(x) = x^{1/2}$$

Since $x^{1/2}$ is a monomial with an exponent, apply the Power Rule.

Step 2: Apply the Power Rule:

$$f'(x) = \frac{1}{2}x^{1/2-1}$$

Step 3: Simplify:

$$f'(x) = \frac{1}{2}x^{-1/2} = \frac{1}{2x^{1/2}} = \frac{1}{2\sqrt{x}}$$

Example 3:

Differentiate the polynomial function $f(x) = 2x^3 - 4x^2 + x - 8$.

Solution:

Step 1: Decide which derivative rule to use.

Use Property 1 for the constant, –8. Use Property 2 to differentiate each of the four terms separately by applying the Power Rule. Use Property 3 for the first three terms.

Step 2: Apply the Power Rule and Properties 1, 2, and 3:

$$f'(x) = 2 \bullet 3 \bullet x^{3-1} - 4 \bullet 2 \bullet x^{2-1} + 1 \bullet x^{1-1} - 0$$

Step 3: Simplify:

$$f'(x) = 6x^2 - 8x^1 + 1x^0$$

$$f'(x) = 6x^2 - 8x + 1$$

CAUTION—Major Mistake Territory!

A common mistake when differentiating multiple terms is to ignore the x-term or the constant. The variable x has an exponent of 1, x^1. In Example 3, the derivative of x is simplified to 1. The derivatives of constants are always 0.

Example 4:

If $y = \frac{1}{x^5}$, find y'.

Solution:

Step 1: Decide which derivative rule to use.

Rewrite the rational function using exponents:

$$y = x^{-5}$$

Since x^{-5} is a monomial with an exponent, apply the Power Rule.

Step 2: Apply the Power Rule:

$$y' = -5 \bullet x^{-5-1}$$

Step 3: Simplify:

$$y' = -5x^{-6} = \frac{-5}{x^6}$$

BRAIN TICKLERS Set # 11

Find the derivative of each of the following expressions, with respect to their variables.

1. $x + 5$
2. $t^3 - 5t^2 + 10t - 8$
3. $\sqrt[3]{x} + \frac{10}{x^7}$
4. $\frac{4}{3\sqrt{v}} - \frac{1}{6v^2} - 2v$

(Answers are on page 81.)

Product Rule

The derivative of a product of two differentiable functions is different from the derivative property for sum and difference. However, if both terms in the product are polynomials, they can be multiplied together, and the Power Rule can be applied.

Example 5:

Find the derivative of $f(x) = (3x - 1)(x + 5)$.

Solution:

Step 1: Decide which derivative rule to use.

Since $f(x)$ is a product of binomials, multiply the binomials, simplify, and apply the Power Rule:

$$f(x) = 3x^2 + 15x - 1x - 5$$

$$f(x) = 3x^2 + 14x - 5$$

Step 2: Apply the Power Rule:

$$f'(x) = 3 \bullet 2 \bullet x^{2-1} + 14 \bullet 1 \bullet x^{1-1} - 0$$

Step 3: Simplify:

$$f'(x) = 6x^1 + 14x^0$$

$$f'(x) = 6x + 14$$

Not all functions will be a product of two polynomial functions. If this is the case, the *Product Rule* for derivatives needs to be applied.

If $f(x) = u(x) \bullet v(x)$, then $f'(x) = u(x) \bullet v'(x) + v(x) \bullet u'(x)$.

1+2=3 MATH TALK!

When applying the Product Rule, think of it as the first times the derivative of the second plus the second times the derivative of the first.

Example 5 revisited:

Find the derivative of $f(x) = (3x - 1)(x + 5)$.

Solution:

Step 1: Decide which derivative rule to use.

Since $f(x)$ is a product of binomials, apply the Product Rule.

Step 2: Apply the Product Rule:

$$f'(x) = \underbrace{(3x - 1)}_{\text{first}}\underbrace{(1)}_{\substack{\text{derivative}\\\text{of second}}} + \underbrace{(x + 5)}_{\text{second}}\underbrace{(3)}_{\substack{\text{derivative}\\\text{of first}}}$$

Step 3: Simplify:

$$f'(x) = (3x - 1) + (3x + 15)$$

$$f'(x) = 6x + 14$$

Quotient Rule

The derivative of a quotient of two differentiable functions is different from the derivative property for sum and difference. However, if the denominator of the function is a monomial, the numerator can be divided by the denominator, and the Power Rule can be applied.

Example 6:

If $f(x) = \dfrac{8x^3 - 1}{x}$, find f'.

Solution:

Step 1: Decide which derivative rule to use.

Since the denominator is the monomial x, divide the numerator by the denominator, simplify using exponents, and apply the Power Rule:

$$f(x) = \frac{8x^3}{x} - \frac{1}{x}$$

$$f(x) = 8x^2 - x^{-1}$$

Step 2: Apply the Power Rule:

$$f' = 8 \cdot 2 \cdot x^{2-1} - (-1) \cdot x^{-1-1}$$

Step 3: Simplify:

$$f' = 16x^1 + x^{-2} = 16x + \frac{1}{x^2}$$

Not all rational expressions can be simplified this way. If there is a polynomial in the denominator, it cannot divide into the numerator, and the *Quotient Rule* for derivatives needs to be applied.

If $f(x) = \dfrac{u(x)}{v(x)}$, then $f'(x) = \dfrac{v(x) \cdot u'(x) - u(x) \cdot v'(x)}{(v(x))^2}$, $v(x) \neq 0$.

1+2=3 MATH TALK!

When applying the Quotient Rule, think of it as:

If $f(x) = \dfrac{hi}{lo}$, then $f'(x) = \dfrac{lo \cdot dhi - hi \cdot dlo}{lo\ lo}$.

Example 7:

Given $y = \dfrac{3x^5 + x^2}{7x - 4}$, find y'.

Solution:

Step 1: Decide which derivative rule to use.

Since y is a rational function with a binomial in the denominator, apply the Quotient Rule.

Step 2: Apply the Quotient Rule:

$$y' = \frac{(7x - 4)(15x^4 + 2x) - (3x^5 + x^2)(7 - 0)}{(7x - 4)^2}$$

Step 3: Simplify:

$$y' = \frac{(7x - 4)(15x^4 + 2x) - 7(3x^5 + x^2)}{(7x - 4)^2}$$

$$y' = \frac{105x^5 + 14x^2 - 60x^4 - 8x - 21x^5 - 7x^2}{49x^2 - 28x - 28x + 16}$$

$$y' = \frac{84x^5 - 60x^4 + 7x^2 - 8x}{49x^2 - 56x + 16}$$

Generally, after completing the Product and Quotient Rules, it is more common *not* to simplify and, instead, to leave your final answer as a series of products.

BRAIN TICKLERS Set # 12

Find the derivative of each of the following expressions, with respect to their variables.

1. $(4x - 5)^2$
2. $\dfrac{6t^4 - 3t^2 + 12}{3t}$
3. $(x^2 - x + 1)(9x + 5)$
4. $\dfrac{2z^{-3} + 5z}{z^6 - 3z^2 + 7}$

(Answers are on page 81.)

Chain Rule

When asked to differentiate a composition of functions (a function within a function), the *Chain Rule* must be applied.

Sometimes when a polynomial function is composed with another polynomial function, the function can be expanded by using the Power Rule instead of the Chain Rule.

Example 8:

Given $y = (3x^5 + x)^2$, find $\frac{dy}{dx}$.

Solution:

Step 1: Decide which derivative rule to use.

Expand the expression and simplify, and then apply the Power Rule:

$$y = (3x^5 + x)(3x^5 + x) = 9x^{10} + 6x^6 + x^2$$

Step 2: Apply the Power Rule:

$$\frac{dy}{dx} = 90x^9 + 36x^5 + 2x$$

1+2=3 MATH TALK!

Example 8 is a composition of functions since the outermost function is a squaring function and the input is a polynomial function.

Not all compositions will be able to expand as nicely, if at all. If that is the case, the Chain Rule needs to be applied.

The Chain Rule can be used to differentiate any composition of differentiable functions.

$$\text{If } f(x) = u(v(x)), \text{ then } f'(x) = u'(v(x)) \bullet v'(x).$$

1+2=3 MATH TALK!

When applying the Chain Rule, think of it as "the derivative of the outer with the same input multiplied by the derivative of the inner."

Example 8 revisited:

Given $y = (3x^5 + x)^2$, find $\frac{dy}{dx}$.

Solution:

Step 1: Decide which derivative rule to use.

Since y is composition of functions with the outer as the squaring function and the input a polynomial function, apply the Chain Rule.

Step 2: Apply the Chain Rule:

$$\frac{dy}{dx} = \underbrace{2 \bullet (3x^5 + x)^{2-1}}_{\text{Power Rule of the outer squaring function}} \bullet \underbrace{(3 \bullet 5 \bullet x^{5-1} + 1 \bullet x^{1-1})}_{\text{Power Rule of the inner function}}$$

Step 3: Simplify:

$$\frac{dy}{dx} = 2(3x^5 + x)(15x^4 + 1)$$

$$\frac{dy}{dx} = (6x^5 + 2x)(15x^4 + 1)$$

$$\frac{dy}{dx} = 90x^9 + 36x^5 + 2x$$

Trigonometric Functions

So far, our discussion of derivatives has centered on polynomial functions or variables that can be raised to exponents. Using the limit definition of a derivative, the derivative rules, or the derivative properties, derivative shortcuts for other functions, including trigonometric functions, can be found.

The derivatives of the six trigonometric functions are shown in the table below.

$f(x)$	$f'(x)$
$\sin(x)$	$\cos(x)$
$\cos(x)$	$-\sin(x)$
$\tan(x)$	$\sec^2(x)$
$\cot(x)$	$-\csc^2(x)$
$\sec(x)$	$\tan(x)\sec(x)$
$\csc(x)$	$-\cot(x)\csc(x)$

Example 9:
If $f(x) = (7x - 6)(\tan x)$, find $f'(x)$.

Solution:

Step 1: Decide which derivative rule to use.

Since $f(x)$ is a product of two different functions (polynomial and trigonometric), apply the Product Rule.

Step 2: Apply the Product Rule:

$$f'(x) = (7x - 6)(\sec^2 x) + (\tan x)(7)$$

Step 3: Simplify:

$$f'(x) = (7x - 6)(\sec^2 x) + 7\tan x$$

Example 10:
Find y' if $y = \sin(x^2 - 5x)$.

Solution:

Step 1: Decide which derivative rule to use.

Since y is a composition of two different functions, with the outer function as the sine function and the input a polynomial function, apply the Chain Rule.

Step 2: Apply the Chain Rule:

$$y' = \underbrace{\cos(x^2 - 5x)}_{\text{derivative of sine with same input}} \bullet \underbrace{(2x - 5)}_{\text{derivative of input}}$$

Step 3: Simplify:

$$y' = (2x - 5)\cos(x^2 - 5x)$$

BRAIN TICKLERS Set # 13

Differentiate each of the following expressions, with respect to their variables.

1. $(7x^3 - 5x + 6)^5$
2. $\dfrac{\tan x}{3x - 1}$
3. $\sin(\sqrt{t})$
4. $(\sqrt{x})(\csc(2x - 1))$

(Answers are on page 81.)

Exponential Functions

The derivative of an exponential function of the form $y = b^x$, $b > 0$ and $b \neq 1$, is $\dfrac{dy}{dx} = b^x \bullet \ln(b)$. This is called the *Exponential Rule.* If we apply this rule to the function $y = e^x$, then $\dfrac{dy}{dx} = e^x \bullet \ln(e) = e^x \bullet 1 = e^x$.

CAUTION—Major Mistake Territory!

Exponential functions are different from polynomial functions although they may look similar. Polynomial functions have variables as the base, raised to numerical exponents. Exponential functions have a numerical value as their base, and their exponent is a variable.

$y = x^2$ is a polynomial function.

$y = 2^x$ is an exponential function.

Example 11:

Differentiate the function $f(x) = e^{2x}$.

Solution:

Step 1: Decide which derivative rule to use.

Rewrite $f(x)$ using exponent properties:

$$f(x) = (e^x)^2$$

Since $f(x)$ is a composition of two different functions, with the outer as the squaring function and the input the exponential function e^x, apply the Chain Rule.

Step 2: Apply the Chain Rule:

$$f'(x) = \underbrace{2 \bullet (e^x)^{2-1}}_{\text{Power Rule of the outer squaring function}} \bullet \underbrace{(e^x)}_{\text{Exponential Rule of the inner function}}$$

Step 3: Simplify:

$$f'(x) = 2(e^x)(e^x) = 2e^{2x}$$

1+2=3 MATH TALK!

The exponential function e^x is very interesting because it is the only function whose derivative is itself. As shown in Example 11, if there is a coefficient in the exponent, the Chain Rule must be applied. In general, the derivative rule for e^x can be extended to include this case:

If $y = e^{ax}$, where constant $a \neq 0$, then $y' = a \cdot e^{ax}$.

Example 12:

Find y' if $y = 3^x$.

Solution:

Step 1: Decide which derivative rule to use.

Since y is an exponential function, apply the Exponential Rule.

Step 2: Apply the Exponential Rule:

$y' = 3^x \cdot \ln 3$

Logarithmic Functions

The derivative of a logarithmic function of the form $y = \log_b x$, $b > 0$, $x > 0$, is $\frac{dy}{dx} = \frac{1}{x \cdot \ln b}$. This is called the *Logarithmic Rule*. When applying this rule to the function $y = \ln x$, since $\ln x = \log_e x$, then $\frac{dy}{dx} = \frac{1}{x \cdot \ln e} = \frac{1}{x \cdot 1} = \frac{1}{x}$.

Example 13:

Find $\frac{dy}{dx}$ if $y = \ln(7x)$.

Solution:

Step 1: Decide which derivative rule to use.

Since y is a composition of two different functions, with the outer as the logarithmic function and the input a polynomial function, $7x$, apply the Chain Rule.

Step 2: Apply the Chain Rule:

$$\frac{dy}{dx} = \underbrace{\frac{1}{7x}}_{\text{Logarithmic Rule of outer function}} \bullet \underbrace{7}_{\text{Power Rule of inner function}}$$

Step 3: Simplify:

$$\frac{dy}{dx} = \frac{7}{7x} = \frac{1}{x}$$

Example 14:

Find $f'(t)$ if $f(t) = 10\log_5 t^3$.

Solution:

Step 1: Decide which derivative rule to use.

Since $f(t)$ is a composition of two different functions, with the outer as the logarithmic function and the input a polynomial function, t^3, apply the Chain Rule.

Step 2: Apply the Chain Rule:

$$f'(t) = 10 \bullet \underbrace{\frac{1}{t^3 \bullet \ln 5}}_{\text{Logarithmic Rule of outer function}} \bullet \underbrace{(3t^2)}_{\text{Power Rule of inner function}}$$

Step 3: Simplify:

$$f'(t) = \frac{30t^2}{t^3 \bullet \ln 5} = \frac{30}{t \bullet \ln 5}$$

To sum it up, remember the following steps when differentiating functions.

Step 1: Decide which derivative rule to use.

Step 2: Apply the rule.

Step 3: Simplify.

Sneaky Derivatives

In Chapter Two, derivatives were evaluated using limit definitions. Derivatives that are represented as limits can also be solved using the derivative shortcuts.

1+2=3 MATH TALK!

To find the derivative function using the limit definition, apply:

$$f'(x) = \lim_{h \to 0} \frac{f(x+h) - f(x)}{h}$$

To evaluate a derivative at a value, $x = a$, using the limit definition, apply:

$$f'(a) = \lim_{h \to 0} \frac{f(a+h) - f(a)}{h} \quad \text{or} \quad f'(a) = \lim_{x \to a} \frac{f(x) - f(a)}{x - a}$$

If a limit has one or more of the following:

- $h \to 0$ or $x \to a$
- two similar functions being subtracted in the numerator
- $x + h$ or $a + h$ in the numerator
- h in the denominator or $x - a$ in the denominator

then most likely the limit is "sneakily" representing the derivative of a function, $f'(x)$ or $f'(a)$. Instead of evaluating the limit, identify the function, take its derivative using derivative shortcut rules, and evaluate the derivative at $x = a$ if necessary.

Example 15:

Evaluate $\lim_{h \to 0} \frac{3(x+h)^2 - 3x^2}{h}$.

Solution:

Since h is approaching 0, two similar functions are being subtracted in the numerator, the input in the first function is $x + h$, and h is in the denominator, this represents the limit definition of $f'(x)$.

Step 1: Identify the function.

Using $f'(x) = \lim_{h \to 0} \frac{f(x+h) - f(x)}{h}$, the function can be identified as the second term in the numerator:

$f(x) = 3x^2$

Step 2: Take the derivative of the function:

$$f'(x) = 6x$$

$$\lim_{h \to 0} \frac{3(x+h)^2 - 3x^2}{h} = 6x$$

Example 16:

Evaluate $\lim_{h \to 0} \frac{(3+h)^4 - 3^4}{h}$.

Solution:

Since h is approaching 0, two similar functions are being subtracted in the numerator, the input in the first function is $3 + h$, and h is in the denominator, this represents the limit definition of $f'(a)$, where $a = 3$.

Step 1: Identify the function.

Using $f'(a) = \lim_{h \to 0} \frac{f(a+h) - f(a)}{h}$, the function can be identified as the second term in the numerator, replacing $a = 3$ with an x:

$$f(x) = x^4$$

Step 2: Take the derivative of the function:

$$f'(x) = 4x^3$$

Step 3: Evaluate the derivative for $x = a = 3$:

$$f'(3) = 4(3)^3 = 108$$

$$\lim_{h \to 0} \frac{(3+h)^4 - 3^4}{h} = 108$$

Example 17:

Evaluate $\lim_{x \to -13} \frac{x^2 - (-13)^2}{x + 13}$.

Solution:

Since x is approaching −13, two similar functions are being subtracted in the numerator, and the denominator is the difference of x and -13, this represents the limit definition of $f'(a)$, where $a = -13$.

Step 1: Identify the function.

Using $f'(a) = \lim_{x \to a} \frac{f(x) - f(a)}{x - a}$, the function can be identified as the second term in the numerator, replacing $a = -13$ with an x:

$$f(x) = x^2$$

Step 2: Take the derivative of the function:

$$f'(x) = 2x$$

Step 3: Evaluate the derivative for $x = a = -13$:

$$f'(-13) = 2(-13) = -26$$

$$\lim_{x \to -13} \frac{x^2 - (-13)^2}{x + 13} = -26$$

BRAIN TICKLERS Set # 14

1. Differentiate with respect to t, $s(t) = \sin(e^{5t})$.
2. Differentiate with respect to x, $y = \ln(x^3 - 4x + 8)$.
3. Evaluate $\lim_{h \to 0} \frac{\sqrt{7 + h} - \sqrt{7}}{h}$.
4. Evaluate $\lim_{h \to 0} \frac{\cos(10 + h) - \cos(10)}{h}$.

(Answers are on page 81.)

Implicit Differentiation

So far, the equations that have been differentiated were all *explicit equations*, meaning the dependent variable, in most cases y, was explicitly expressed in terms of the independent variable, in most cases x. Sometimes it is not possible to isolate the dependent variable. For example, given the equation $x^2 - y^2 + 4xy = 10$, it is impossible to solve for y with respect to x. The derivative of this type of equation can still be found using a method called *implicit differentiation*.

Implicit differentiation is *painless*. There are four steps to finding the derivative of a function.

Step 1: Differentiate both sides of the equation with respect to x (or with respect to the independent variable).

Step 2: Isolate the derivative, such as $\frac{dy}{dx}$, to one side of the equation.

Step 3: Factor out the derivative, such as $\frac{dy}{dx}$, if necessary.

Step 4: Solve for the derivative, such as $\frac{dy}{dx}$.

1+2=3 MATH TALK!

Now that derivatives can be found implicitly, it is important that the variable the equation is being differentiated with respect to is either stated or notated. If the equation is being differentiated with respect to x, the usual shortcut rules are applied to x. However, differentiating any other variable must include a derivative placeholder since what that variable represents is not explicitly known.

For example, the derivative of y^5 with respect to x uses the Chain Rule since the outer function is a fifth power and the input is y, which is treated as a function of x:

$$\frac{d}{dx}(y^5) = \underbrace{5y^4}_{\text{Power Rule of outer function}} \bullet \underbrace{\frac{dy}{dx}}_{\text{Derivative of inner function}} = 5y^4\frac{dy}{dx}$$

Example 18:

Differentiate $x^2 - y^2 + 4xy = 10$ with respect to x.

Solution:

Step 1: Differentiate both sides of the equation with respect to x:

$$\frac{d}{dx}(x^2 - y^2 + 4xy) = \frac{d}{dx}(10)$$

Using Property 2, each term will be differentiated with respect to x using the respective derivative shortcuts. If the variable is not x, the derivative rule will still be applied, but a placeholder will be left notating the derivative of the input:

$$\underbrace{2x}_{\text{Power Rule}} - \underbrace{2y\frac{dy}{dx}}_{\text{Power Rule}} + \underbrace{(4x)\left(\frac{dy}{dx}\right) + (y)(4)}_{\text{Product Rule}} = \underbrace{0}_{\text{Property 1}}$$

Step 2: Isolate the derivative, $\frac{dy}{dx}$, to one side of the equation:

$$-2y\frac{dy}{dx} + (4x)\left(\frac{dy}{dx}\right) = -2x - 4y$$

Step 3: Factor out the derivative $\frac{dy}{dx}$:

$$\frac{dy}{dx}(-2y + 4x) = -2x - 4y$$

Step 4: Solve for the derivative $\frac{dy}{dx}$:

$$\frac{dy}{dx} = \frac{-2x - 4y}{-2y + 4x}$$

Example 19:

If $\cos(xy) = y$, find $\frac{dy}{dx}$.

Solution:

Step 1: Differentiate both sides of the equation with respect to x:

$$\frac{d}{dx}(\cos(xy)) = \frac{d}{dx}(y)$$

$$-\sin(xy) \bullet \left(x \bullet \frac{dy}{dx} + y \bullet 1\right) = \frac{dy}{dx}$$

$$-x\sin(xy)\frac{dy}{dx} - y\sin(xy) = \frac{dy}{dx}$$

Step 2: Isolate the derivative, $\frac{dy}{dx}$, to one side of the equation:

$$-x\sin(xy)\frac{dy}{dx} - \frac{dy}{dx} = y\sin(xy)$$

Step 3: Factor out the derivative $\frac{dy}{dx}$:

$$\frac{dy}{dx}(-x\sin(xy) - 1) = y\sin(xy)$$

Step 4: Solve for the derivative $\frac{dy}{dx}$:

$$\frac{dy}{dx} = \frac{y\sin(xy)}{-x\sin(xy) - 1}$$

The second derivative can also be found with implicit differentiation. The process is the same as before: find the first derivative and then differentiate a second time. In the second differentiation, there will be $\frac{dy}{dx}$ terms where the first derivative is substituted.

Example 20:

Find the second derivative of $x^3 + y^3 = 1$ with respect to x.

Solution:

Step 1: Follow the steps to find $\frac{dy}{dx}$:

$$3x^2 + 3y^2\frac{dy}{dx} = 0$$

$$\frac{dy}{dx} = \frac{-3x^2}{3y^2} = \frac{-x^2}{y^2}$$

Step 2: Differentiate again to find $\frac{d^2y}{dx^2}$. Use the Quotient Rule:

$$\frac{d^2y}{dx^2} = \frac{(y^2)(-2x) - (-x^2)\left(2y\frac{dy}{dx}\right)}{(y^2)^2}$$

Step 3: Substitute for $\frac{dy}{dx}$ and simplify:

$$\frac{d^2y}{dx^2} = \frac{-2xy^2 + 2x^2y\left(\frac{-x^2}{y^2}\right)}{y^4} = \frac{-2xy^2 + \frac{-2x^4}{y}}{y^4}$$

$$= \frac{-2xy^2 + \frac{-2x^4}{y}}{y^4} \cdot \frac{y}{y} = \frac{-2xy^3 - 2x^4}{y^5}$$

BRAIN TICKLERS Set # 15

1. If $y^3 - x = x^2 + 3y$, find $\frac{dy}{dx}$.
2. Differentiate, with respect to x, $y^3 + 2xy = 5$.
3. If $\frac{dy}{dx} = 2x - 3y + 7$, find $\frac{d^2y}{dx^2}$ in terms of x and y.
4. If $x^2 + y^2 + 5y = 11$, find $\frac{d^2y}{dx^2}$ in terms of x and y.

(Answers are on page 82.)

Logarithmic Differentiation

Functions of the form $f(x) = u^v$, where u and v are nonconstant functions of x, can be differentiated using *logarithmic differentiation.*

Step 1: Take the natural log of both sides of the equation.

Step 2: Using logarithmic properties, bring down the power in front of the natural log.

Step 3: Differentiate both sides of the equation using the Logarithmic Rule and the Product Rule.

Step 4: Solve for $\frac{dy}{dx}$.

CAUTION—Major Mistake Territory!

When asked to differentiate a function that has a power, only use logarithmic differentiation when the variable is in both the base *and* the power. If the base or power is a constant, other shortcuts should be applied.

Example 21:

Differentiate $y = (x^2 + 1)^{\sin x}$.

Solution:

Step 1: Take the natural log of both sides of the equation:

$$\ln(y) = \ln((x^2 + 1)^{\sin x})$$

Step 2: Using logarithmic properties, bring the power in front of the natural log:

$$\ln(y) = \sin x \bullet \ln(x^2 + 1)$$

Step 3: Differentiate both sides of the equation using the Logarithmic Rule and the Product Rule:

$$\frac{1}{y} \bullet \frac{dy}{dx} = (\sin x)\left(\frac{1}{x^2 + 1} \bullet 2x\right) + (\ln(x^2 + 1))(\cos x)$$

Step 4: Solve for $\frac{dy}{dx}$, replacing y with the information given in the problem:

$$\frac{dy}{dx} = y\left[(\sin x)\left(\frac{1}{x^2 + 1} \bullet 2x\right) + (\ln(x^2 + 1))(\cos x)\right]$$

$$\frac{dy}{dx} = (x^2 + 1)^{\sin x}\left[\frac{2x \sin x}{x^2 + 1} + \cos x \ln(x^2 + 1)\right]$$

Inverse Functions

One way to find the derivative of the inverse of a function would be to find the equation of the inverse and then differentiate. Often it is not possible to find an inverse equation. Instead, the following rule can be used to calculate the derivative of an inverse:

$$\frac{d}{dx}(f^{-1}(x)) = \frac{1}{f'(f^{-1}(x))}$$

CAUTION—Major Mistake Territory!

A common mistake is to confuse the notation of $f^{-1}(x)$ with the exponent notation of x^{-1}. The function notation $f^{-1}(x)$ represents an inverse function, where the input (domain) and output (range) are "switched" from the original $f(x)$. The exponent notation x^{-1} represents the reciprocal $\frac{1}{x}$.

Example 22:

If $f(x) = x^3 + 1$, evaluate $\frac{d}{dx}(f^{-1}(9))$.

Solution:

Use the formula $\frac{d}{dx}(f^{-1}(9)) = \frac{1}{f'(f^{-1}(9))}$. First find the value of $f^{-1}(9)$. Instead of finding an equation for $f^{-1}(x)$ and substituting in $x = 9$, set $f(x) = 9$. This is because the input of the inverse function, $f^{-1}(x)$, is also the output of the function, $f(x)$. Then solve for x:

$$f(x) = 9$$

$$x^3 + 1 = 9$$

$$x = 2$$

Therefore, $f^{-1}(9) = 2$.

Thus, $\frac{d}{dx}(f^{-1}(9)) = \frac{1}{f'(2)}$.

To evaluate $f'(2)$, differentiate $f(x)$ and substitute in $x = 2$:

$$f'(x) = 3x^2$$

$$f'(2) = 3(2)^2 = 12$$

The solution is $\frac{d}{dx}(f^{-1}(9)) = \frac{1}{12}$.

Example 23:

Given inverse functions $f(x)$ and $g(x)$, where $f(1) = 5$ and $f'(1) = 2$, what is $g'(5)$?

Solution:

Since $g(x) = f^{-1}(x)$, then $g'(5) = \dfrac{1}{f'(f^{-1}(5))}$. Since it was given that $f(1) = 5$, then $f^{-1}(5) = 1$:

$$g'(5) = \frac{1}{f'(f^{-1}(5))} = \frac{1}{f'(1)} = \frac{1}{2}$$

Example 24:

Suppose $f(x)$ and $g(x)$ and their inverses $f^{-1}(x)$ and $g^{-1}(x)$ are differentiable functions. Let the values of $f(x)$, $g(x)$ and the derivatives $f'(x)$ and $g'(x)$ at $x = 4$ and $x = 5$ be given by the table below.

x	f (x)	g(x)	f′(x)	g′(x)
4	1	5	2	6
5	5	3	4	8

Find the derivative of $g^{-1}(x)$ at $x = 5$.

Solution:

Use the formula $\dfrac{d}{dx}(g^{-1}(5)) = \dfrac{1}{g'(g^{-1}(5))}$. First find the value of $g^{-1}(5)$. Since there are no columns labeled $g^{-1}(x)$ to look up $x = 5$, set $g(x) = 5$. This is because the input of the inverse function, $g^{-1}(x)$, is also the output of the function, $g(x)$. Then solve for x using the information in the table:

$$g(x) = 5$$

$$x = 4$$

Therefore, $g^{-1}(5) = 4$. Use this information to solve the problem:

$$\frac{d}{dx}(g^{-1}(5)) = \frac{1}{g'(g^{-1}(5))} = \frac{1}{g'(4)} = \frac{1}{6}$$

BRAIN TICKLERS Set # 16

1. If $y = (1+x)^x$, find $\frac{dy}{dx}$.
2. Find the derivative with respect to x for $y = x^{e^x}$.
3. Suppose $f(x) = x^3 + 2x + 1$. What is $\frac{d}{dx}(f^{-1}(4))$?
4. Suppose $f^{-1}(2) = 3$ and $\frac{d}{dx}(f^{-1}(2)) = 6$. What is $f'(3)$?

(Answers are on page 82.)

BRAIN TICKLERS—THE ANSWERS

Set # 11, page 59

1. 1
2. $3t^2 - 10t + 10$
3. $\frac{1}{3}x^{-2/3} - 70x^{-8} = \frac{1}{3\sqrt[3]{x^2}} - \frac{70}{x^8}$
4. $-\frac{2}{3}v^{-3/2} + \frac{1}{3}v^{-3} - 2 = -\frac{2}{3\sqrt{v^3}} + \frac{1}{3v^3} - 2$

Set # 12, page 62

1. $32x - 40$
2. $6t^2 - 1 - 4t^{-2} = 6t^2 - 1 - \frac{4}{t^2}$
3. $(x^2 - x + 1)(9) + (9x + 5)(2x - 1) = 27x^2 - 8x + 4$
4. $\frac{(z^6 - 3z^2 + 7)(-6z^{-4} + 5) - (2z^{-3} + 5z)(6z^5 - 6z)}{(z^6 - 3z^2 + 7)^2}$

Set # 13, page 66

1. $5(7x^3 - 5x + 6)^4(21x^2 - 5)$
2. $\frac{(3x - 1)(\sec^2 x) - (\tan x)(3)}{(3x - 1)^2}$
3. $\cos(\sqrt{t}) \bullet \frac{1}{2}t^{-1/2} = \frac{\cos(\sqrt{t})}{2\sqrt{t}}$
4. $(\sqrt{x})(-\cot(2x - 1)\csc(2x - 1) \bullet (2)) + (\csc(2x - 1))\left(\frac{1}{2}x^{-1/2}\right)$

Set # 14, page 72

1. $s'(t) = \cos(e^{5t}) \bullet (5e^{5t})$
2. $\frac{dy}{dx} = \frac{1}{x^3 - 4x + 8} \bullet (3x^2 - 4)$
3. $\frac{1}{2\sqrt{7}}$
4. $-\sin(10)$

Set # 15, page 76

1. $\dfrac{dy}{dx} = \dfrac{2x+1}{3y^2-3}$

2. $\dfrac{dy}{dx} = \dfrac{-2y}{3y^2+2x}$

3. $\dfrac{d^2y}{dx^2} = -6x+9y-19$

4. $\dfrac{d^2y}{dx^2} = \dfrac{(2y+5)(-2)-(-2x)\left(2\dfrac{dy}{dx}\right)}{(2y+5)^2}$

$$= \frac{-4y-10+2x^2\left(\dfrac{-2x}{2y+5}\right)}{(2y+5)^2} = \frac{-4y-10-\dfrac{4x^3}{(2y+5)}}{(2y+5)^2}$$

Set # 16, page 80

1. $\dfrac{dy}{dx} = (1+x)^x\left[\dfrac{x}{(1+x)} + \ln(1+x)\right]$

2. $x^{e^x}\left(\dfrac{e^x}{x} + e^x \ln x\right)$

3. $\dfrac{1}{5}$

4. $\dfrac{1}{6}$

Chapter 4

Applications of Derivatives

Knowing how to calculate the derivative of a function and what it represents leads to some very useful applications of derivatives.

Equation of a Tangent Line

Discovering that the derivative of a function at a value represents the slope of the tangent line can be extended further to writing the equation of a tangent line to a curve at a point. There are a few ways to write the equation of a line, including the **slope-intercept form,** $y = mx + b$, and the **point-slope form,** $y - y_1 = m(x - x_1)$. Typically, you will be given the equation of the function and the x-value at which the tangent line and the curve intersect. With this given information, using the point-slope form to write the equation of a tangent line will be more convenient.

Writing the equation of a tangent line is *painless*. There are three steps to follow.

Step 1: Find the coordinates of the point of tangency; these are the x- and y-coordinates of the point of intersection of the tangent line and the curve.

Step 2: Find the slope of the tangent line by calculating the derivative at the point of tangency.

Step 3: Substitute the point of tangency and the slope of the tangent into the point-slope form of a line.

Example 1:

Given the function $y = x^3$, write the equation of the tangent line to the curve at $x = 2$.

Solution:

Step 1: Find the coordinates of the point of tangency.

Given $y = x^3$ and $x = 2$:

$$y = (2)^3 = 8$$

The point of tangency is (2, 8).

Step 2: Find the slope of the tangent line.

Calculate the derivative:

$$\frac{dy}{dx} = 3x^2$$

Evaluate the derivative at $x = 2$:

$$\frac{dy}{dx} = 3(2)^2 = 12$$

This represents the slope of the tangent line at $x = 2$.

Step 3: Use point-slope form, $y - y_1 = m(x - x_1)$, to find the equation:

Point: $(x_1, y_1) = (2, 8)$

Slope: $m = 12$

$y - 8 = 12(x - 2)$ or $y = 12x - 16$

PAINLESS TIP

Unless otherwise stated, the equation for a tangent line can be left in point-slope form. Since the point of tangency will very rarely be a *y*-intercept, you should always use the point-slope form to write the equation of the line since finding the coordinates of the tangent point and the tangent slope can be obtained using the given function and its derivative.

Example 2:

Find the equation of the tangent to the curve $f(x) = 6\sqrt{x} - x$ at $x = 1$.

Solution:

Step 1: Find the coordinates of the point of tangency.

Given $f(x) = 6\sqrt{x} - x$ and $x = 1$:

$$f(1) = 6\sqrt{1} - 1 = 5$$

The point of tangency is $(1, 5)$.

Step 2: Find the slope of the tangent line.

Calculate the derivative:

$$f'(x) = 3x^{-1/2} - 1$$

Evaluate the derivative at $x = 1$:

$$f'(1) = 3(1)^{-1/2} - 1 = 2$$

Step 3: Use point-slope form, $y - y_1 = m(x - x_1)$, to find the equation:

Point: $(x_1, y_1) = (1, 5)$

Slope: $m = 2$

$y - 5 = 2(x - 1)$

Example 3:

If $f(1) = 5$ and $f'(1) = 6$, write the equation of the tangent line to the curve at $x = 1$.

Solution:

Step 1: Find the coordinates of the point of tangency.

Given $f(1) = 5$, the point of tangency is $(1, 5)$.

Step 2: Find the slope of the tangent line.

Given $f'(1) = 6$, the slope of the tangent line is 6.

Step 3: Use point-slope form, $y - y_1 = m(x - x_1)$, to find the equation:

Point: $(x_1, y_1) = (1, 5)$

Slope: $m = 6$

$y - 5 = 6(x - 1)$

Equation of a Normal Line

A normal to a curve is a line that is perpendicular to the tangent at the point of tangency.

> **1+2=3 MATH TALK!**
>
> For two lines to be perpendicular, their slopes are negative reciprocals. To find the equation of a normal line, follow the same steps to find the equation of a tangent line. The only change is to use the negative reciprocal of the derivative as the slope of the normal line.

Example 4:

Find the equation of the normal to $y = \frac{7}{\sqrt{x}}$ at the point where $x = 4$.

Solution:

Step 1: Find the coordinates of the point of tangency.

Given $y = \frac{7}{\sqrt{x}}$ and $x = 4$:

$y = \frac{7}{\sqrt{4}} = \frac{7}{2}$

The point of tangency is $\left(4, \frac{7}{2}\right)$.

Step 2: Find the slope of the normal line.

Calculate the derivative by rewriting the equation as $y = 7x^{-1/2}$:

$$\frac{dy}{dx} = -\frac{7}{2}x^{-3/2}$$

Find the derivative at $x = 4$:

$$\frac{dy}{dx} = -\frac{7}{2}(4)^{-3/2} = -\frac{7}{16}$$

This represents the slope of the tangent line at $x = 4$. Therefore, the slope of the normal line is $+\frac{16}{7}$.

Step 3: Use point-slope form, $y - y_1 = m(x - x_1)$, to find the equation:

Point: $(x_1, y_1) = \left(4, \frac{7}{2}\right)$

Slope: $m = \frac{16}{7}$

$$y - \frac{7}{2} = \frac{16}{7}(x - 4)$$

Example 5:

Find an equation for the line that is normal to $y = \frac{1-x}{1+x}$ at the point where $x = 2$.

Solution:

Step 1: Find the coordinates of the point of tangency.

Given $y = \frac{1-x}{1+x}$ and $x = 2$:

$$y = \frac{1-2}{1+2} = \frac{-1}{3}$$

The point of tangency is $\left(2, -\frac{1}{3}\right)$.

Step 2: Find the slope of the normal line.

Calculate the derivative using the Quotient Rule:

$$\frac{dy}{dx} = \frac{(1+x)(-1)-(1-x)(1)}{(1+x)^2}$$

Calculate the derivative at $x = 2$:

$$\frac{dy}{dx} = \frac{(1+2)(-1)-(1-2)(1)}{(1+2)^2} = -\frac{2}{9}$$

This represents the slope of the tangent line at $x = 2$. Therefore, the slope of the normal line is $+\frac{9}{2}$.

Step 3: Use point-slope form, $y - y_1 = m(x - x_1)$, to find the equation:

Point: $(x_1, y_1) = \left(2, -\frac{1}{3}\right)$

Slope: $m = \frac{9}{2}$

$$y + \frac{1}{3} = \frac{9}{2}(x-2)$$

Local Linear Approximation

One application of the equation of a tangent line is to make approximations for other x-values on the curve. Sometimes evaluating functions can be challenging, whereas evaluating a linear function is much easier.

In general, if a function f is differentiable at an x-value, a sufficiently magnified portion of the graph of f centered at the point $(x, f(x))$ takes on the appearance of a straight line segment. For this reason, a function that is differentiable at x is sometimes said to be locally linear at x.

1+2=3 **MATH TALK!**

If you zoom in on the graph of a curve and its tangent line at the point of tangency, they both appear linear and indistinguishable. For example, consider the graph of the function $f(x) = x^2 + 1$ and its tangent line $y = 1$ at the point (0, 1) shown below.

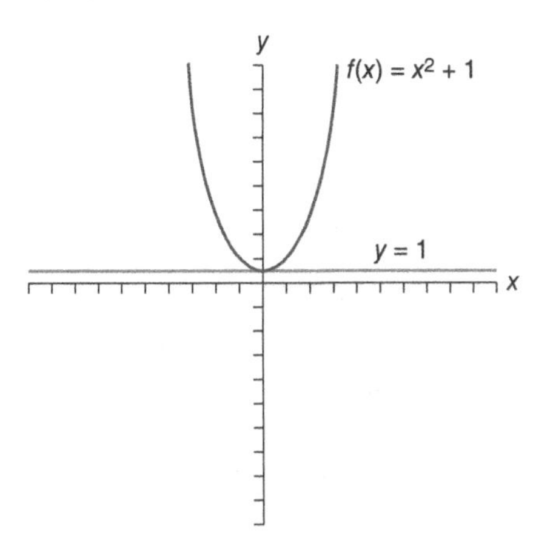

When zooming in on the point (0, 1), the graph of $f(x)$ appears to be linear and takes on the appearance of the tangent line $y = 1$.

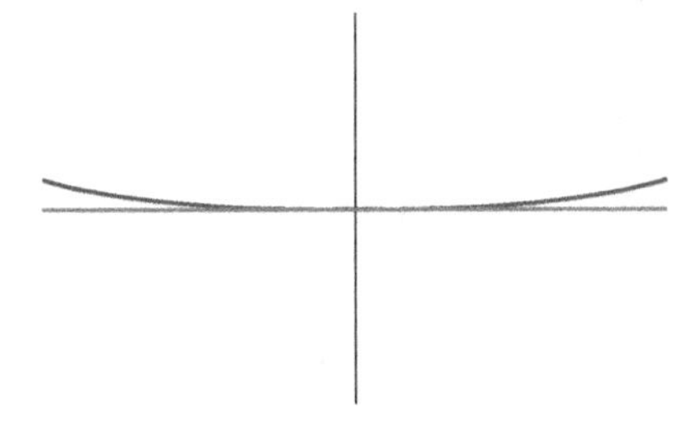

This allows an x-value that is close to the point of tangency to be substituted into the equation of the tangent line and be used as an approximation for the actual function value.

Local linear approximation is *painless*. There are three steps to follow.

Step 1: Identify the function that is to be approximated and an x-value that is close to the given value that is to be approximated.

Step 2: Find the equation of the tangent line of the function at the chosen x-value.

Step 3: Substitute the given x-value into the tangent line equation and solve for y. This y-value is the approximation value and is either an overestimate or an underestimate of the actual value on the curve.

Example 6:

Use local linear approximation to approximate $\sqrt{4.1}$.

Solution:

Step 1: Identify the function that is to be approximated and an x-value that is close to the given value.

The function being used is the square root function, $f(x) = \sqrt{x}$. The given x-value is 4.1. A close x-value that can be easily evaluated is $x = 4$.

Step 2: Find the equation of the tangent line of the function at the chosen x-value.

To find the point of tangency, evaluate $f(4) = \sqrt{4} = 2$. The point of tangency is (4, 2). To find the slope of the tangent line, evaluate the derivative at $x = 4$:

$$f'(x) = \frac{1}{2}(x)^{-1/2}$$

$$f'(4) = \frac{1}{2}(4)^{-1/2} = \frac{1}{4}$$

The equation of the tangent line is $y - 2 = \frac{1}{4}(x - 4)$.

Step 3: Substitute the given x-value into the tangent line equation and solve for y:

$$y - 2 = \frac{1}{4}((4.1) - 4)$$

$$y - 2 = \frac{1}{4}(0.1)$$

$$y = 2.025$$

Therefore, $\sqrt{4.1} \approx 2.025$.

Typing $\sqrt{4.1}$ into the graphing calculator yields the value 2.024845672. This is very close to the approximated value of 2.025 found using the equation of the tangent line in Example 6. The approximation is greater than the actual value because of the curvature of the function. As shown below, the tangent line is graphed above the function due to the function's curvature.

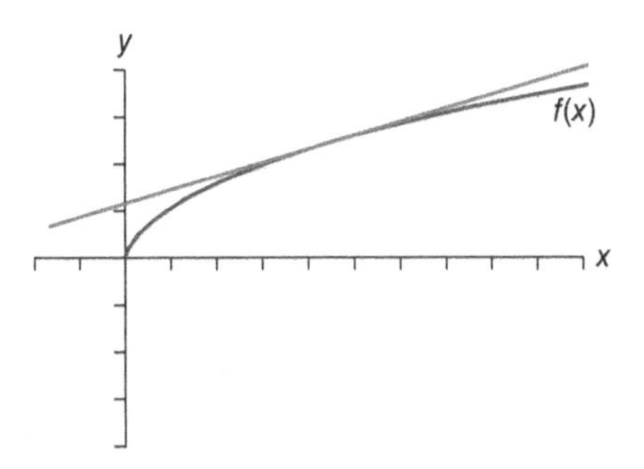

Example 7:

Given $f(x) = \sqrt{9 + \sin x}$, find the approximate value at $x = 0.15$ obtained from the tangent to the graph at $x = 0$.

Solution:

Step 1: Identify the function that is to be approximated and an x-value that is close to the given value.

The function being used is given, $f(x) = \sqrt{9 + \sin x}$. The given x-value is 0.15. An x-value close to the given that can be easily evaluated is $x = 0$, which is also given.

Step 2: Find the equation of the tangent line of the function at the chosen x-value.

To find the point of tangency, evaluate $f(0)$:

$$f(0) = \sqrt{9 + \sin(0)} = \sqrt{9 + 0} = 3$$

The point of tangency is $(0, 3)$. To find the slope of the tangent line, evaluate the derivative at $x = 0$:

$$f'(x) = \frac{1}{2}(9 + \sin x)^{-1/2} \bullet \cos x$$

$$f'(0) = \frac{1}{2}(9 + \sin(0))^{-1/2} \bullet \cos(0) = \frac{1}{6} \bullet 1 = \frac{1}{6}$$

The equation of the tangent line is $y - 3 = \frac{1}{6}(x - 0)$.

Step 3: Substitute the given x-value into the tangent line equation and solve for y:

$$y - 3 = \frac{1}{6}((0.15) - 0)$$

$$y - 3 = \frac{1}{6}(0.15)$$

$$y = 3.025$$

Therefore, $f(0.15) \approx 3.025$.

BRAIN TICKLERS Set # 17

1. Write the equation of the line tangent to the graph of $y = x^4$ at $x = -1$.
2. Write the equation of the line tangent to the graph of $y = \ln(x)$ at $x = e^3$.
3. Write the equation of the line normal to the graph of $f(x) = (1 - 2x)^3$ at the point $(1, -1)$.
4. Use the equation of the line tangent to the graph of $y = \ln\left(\frac{x}{2}\right)$ at $x = 2$ to approximate the value at $x = 2.1$.

(Answers are on page 153.)

L'Hopital's Rule

An application of local linear approximation allows indeterminate limits to be evaluated using their derivatives in certain cases. This is known as *L'Hopital's Rule.*

L'Hopital's Rule Case 1

If $\lim_{x \to a} f(x) = 0$ and $\lim_{x \to a} g(x) = 0$, then $\lim_{x \to a} \frac{f(x)}{g(x)} = \lim_{x \to a} \frac{f'(x)}{g'(x)}$.

1+2=3 MATH TALK!

When evaluating a limit that produces the indeterminate form $\frac{0}{0}$, L'Hopital's Rule can be applied. To evaluate the limit, take the derivative of the numerator and separately take the derivative of the denominator. Then evaluate the limit. If the indeterminate form $\frac{0}{0}$ appears again, apply L'Hopital's Rule again as needed.

Example 8:

Evaluate $\lim_{x \to 0} \frac{\sin x}{x}$.

Solution:

Since $\lim_{x \to 0} \sin x = 0$ and $\lim_{x \to 0} x = 0$, then by L'Hopital's Rule,

$$\lim_{x \to 0} \frac{\sin x}{x} = \lim_{x \to 0} \frac{\cos x}{1} = \lim_{x \to 0} \cos x = 1.$$

Example 9:

Evaluate $\lim_{x \to 0} \frac{e^x - 1}{x^3}$.

Solution:

Since $\lim_{x \to 0} e^x - 1 = 0$ and $\lim_{x \to 0} x^3 = 0$, then by L'Hopital's Rule,

$$\lim_{x \to 0} \frac{e^x - 1}{x^3} = \lim_{x \to 0} \frac{e^x}{3x^2} = \frac{1}{0}, \text{ which does not exist.}$$

L'Hopital's Rule Case 2

If $\lim_{x \to a} f(x) = \pm\infty$ and $\lim_{x \to a} g(x) = \pm\infty$, then $\lim_{x \to a} \frac{f(x)}{g(x)} = \lim_{x \to a} \frac{f'(x)}{g'(x)}$.

Note: The signs of the infinities do not have to match; if it is some ratio of infinity, L'Hopital's Rule may be applied.

Example 10:

Evaluate $\lim\limits_{x \to \frac{\pi}{2}^-} \frac{\sec x}{1 + \tan x}$.

Solution:

Since $\lim\limits_{x \to \frac{\pi}{2}^-} \sec x = \infty$ and $\lim\limits_{x \to \frac{\pi}{2}^-} 1 + \tan x = \infty$, then by L'Hopital's Rule, $\lim\limits_{x \to \frac{\pi}{2}^-} \frac{\sec x}{1 + \tan x} = \lim\limits_{x \to \frac{\pi}{2}^-} \frac{\sec x \tan x}{\sec^2 x}$. By simplifying and using trigonometric substitutions:

$$\lim_{x \to \frac{\pi}{2}^-} \frac{\sec x \tan x}{\sec^2 x} = \lim_{x \to \frac{\pi}{2}^-} \frac{\tan x}{\sec x} = \lim_{x \to \frac{\pi}{2}^-} \frac{\frac{\sin x}{\cos x}}{\frac{1}{\cos x}} = \lim_{x \to \frac{\pi}{2}^-} \sin x = 1$$

CAUTION—Major Mistake Territory!

Although L'Hopital's Rule is more convenient than some of the limit techniques discussed in Chapter One, it may be applied only if a limit is either one of the two indeterminate forms: $\frac{0}{0}$ or $\frac{\infty}{\infty}$. You must check for these two indeterminate forms first. Otherwise, applying L'Hopital's Rule may result in an incorrect answer. If a limit simplifies to $\frac{1}{0}$ or ∞, there is no need for L'Hopital's Rule since the limit does not exist.

Example 11:

Evaluate $\lim\limits_{x \to 0} \frac{\cos x}{2x}$.

Solution:

Since $\lim\limits_{x \to 0} \cos x = 1$ and $\lim\limits_{x \to 0} 2x = 0$, L'Hopital's Rule cannot be applied. The $\lim\limits_{x \to 0} \frac{\cos x}{2x} = \frac{1}{0}$, which means the limit does not exist.

Sometimes when other indeterminate forms appear, the expression can be manipulated so that L'Hopital's Rule can be applied.

Example 12:

Evaluate $\lim_{x\to\infty}\left(x\sin\left(\frac{1}{x}\right)\right)$.

Solution:

When evaluating the behavior of each part of the expression, the limit simplifies to $\infty \bullet 0$. This is an indeterminate form. In order to apply L'Hopital's Rule, the expression must be manipulated to create a ratio. Dividing by the reciprocal of one of the terms does not change the expression and creates the necessary ratio.

Dividing by the reciprocal of x:

$$\lim_{x\to\infty}\left(x\sin\left(\frac{1}{x}\right)\right) = \lim_{x\to\infty}\frac{\sin\left(\frac{1}{x}\right)}{\frac{1}{x}}$$

Since $\lim_{x\to\infty}\sin\left(\frac{1}{x}\right) = 0$ and $\lim_{x\to\infty}\frac{1}{x} = 0$, by L'Hopital's Rule:

$$\lim_{x\to\infty}\frac{\sin\left(\frac{1}{x}\right)}{\frac{1}{x}} = \lim_{x\to\infty}\frac{\cos\left(\frac{1}{x}\right)\bullet -1x^{-2}}{-1x^{-2}}$$

Dividing out the common factor:

$$\lim_{x\to\infty}\frac{\cos\left(\frac{1}{x}\right)\bullet -1x^{-2}}{-1x^{-2}} = \lim_{x\to\infty}\cos\left(\frac{1}{x}\right) = 1$$

Example 13:

Evaluate $\lim_{x\to 1}\left(\frac{1}{\ln x} - \frac{1}{x-1}\right)$.

Solution:

When evaluating the behavior of each part of the expression, the limit simplifies to $\infty - \infty$. This is an indeterminate form. In order to apply L'Hopital's Rule, the expression must be manipulated to create a ratio. Combine the fractions by finding a common denonminator to write the limit as:

$$\lim_{x\to 1}\left(\frac{1}{\ln x}-\frac{1}{x-1}\right)=\lim_{x\to 1}\frac{x-1-\ln x}{(\ln x)(x-1)}$$

Since $\lim_{x\to 1}(x-1-\ln x)=0$ and $\lim_{x\to 1}(\ln x \bullet (x-1))=0$, by L'Hopital's Rule:

$$\lim_{x\to 1}\frac{x-1-\ln x}{(\ln x)(x-1)}=\lim_{x\to 1}\frac{1-\frac{1}{x}}{\ln x(1)+(x-1)\frac{1}{x}}$$

Since $\lim_{x\to 1}\left(1-\frac{1}{x}\right)=0$ and $\lim_{x\to 1}\left(\ln x+(x-1)\frac{1}{x}\right)=0$, by L'Hopital's Rule:

$$\lim_{x\to 1}\frac{1-\frac{1}{x}}{\ln x(1)+(x-1)\frac{1}{x}}=\lim_{x\to 1}\frac{x^{-2}}{\frac{1}{x}+x^{-2}}=\frac{1}{2}$$

Other indeterminate forms involve exponents, such as 1^{∞}, 0^{0}, ∞^{0}. In order to apply L'Hopital's Rule, logarithms are used to manipulate these expressions.

Evaluating these indeterminate forms is *painless* and involves four steps.

Step 1: Let y equal the expression being evaluated.

Step 2: Take the limit of each side and then the natural log of each side.

Step 3: Use logarithmic properties to bring the exponent in front of the log.

Step 4: Evaluate the limit, and solve for the limit of y.

Example 14:

Evaluate $\lim_{x\to\infty}\left(1+\frac{1}{x}\right)^{x}$.

Solution:

When evaluating the behavior of each part of the expression, the limit simplifies to 1^{∞}. This is an indeterminate form.

Step 1: Let y equal the expression being evaluated:

$$y = \left(1 + \frac{1}{x}\right)^x$$

Step 2: Take the limit of each side and then the natural log of each side:

$$\lim_{x \to \infty} y = \lim_{x \to \infty}\left(1 + \frac{1}{x}\right)^x$$

$$\ln\left(\lim_{x \to \infty} y\right) = \ln\left(\lim_{x \to \infty}\left(1 + \frac{1}{x}\right)^x\right)$$

Step 3: Use logarithmic properties to bring the exponent in front of the natural log:

$$\ln\left(\lim_{x \to \infty} y\right) = x\ln\left(\lim_{x \to \infty}\left(1 + \frac{1}{x}\right)\right)$$

Step 4: Evaluate the limit, and solve for the limit of y:

$$\ln\left(\lim_{x \to \infty} y\right) = \lim_{x \to \infty} x\ln\left(1 + \frac{1}{x}\right)$$

$$\ln\left(\lim_{x \to \infty} y\right) = \lim_{x \to \infty}\frac{\ln\left(1 + \frac{1}{x}\right)}{\frac{1}{x}}$$

By L'Hopital's Rule:

$$\ln\left(\lim_{x \to \infty} y\right) = \lim_{x \to \infty}\frac{\frac{1}{1 + \frac{1}{x}} \bullet -x^{-2}}{-x^{-2}}$$

$$\ln\left(\lim_{x \to \infty} y\right) = \lim_{x \to \infty}\frac{1}{1 + \frac{1}{x}}$$

$$\ln\left(\lim_{x \to \infty} y\right) = 1$$

Rewrite in exponential form:

$$\lim_{x \to \infty} y = e^1 = e$$

Therefore, $\lim_{x \to \infty}\left(1 + \frac{1}{x}\right)^x = e.$

Example 15:

Evaluate $\lim_{x \to \infty} x^{1/x}$.

Solution:

When evaluating the behavior of each part of the expression, the limit simplifies to ∞^0. This is an indeterminate form.

Step 1: Let y equal the expression being evaluated:

$$y = x^{1/x}$$

Step 2: Take the limit of each side and then the natural log of each side:

$$\lim_{x \to \infty} y = \lim_{x \to \infty} x^{1/x}$$

$$\ln\left(\lim_{x \to \infty} y\right) = \ln\left(\lim_{x \to \infty} x^{1/x}\right)$$

Step 3: Use logarithmic properties to bring the exponent in front of the natural log:

$$\ln\left(\lim_{x \to \infty} y\right) = \frac{1}{x}\ln\left(\lim_{x \to \infty} x\right)$$

Step 4: Evaluate the limit, and solve for the limit of y:

$$\ln\left(\lim_{x \to \infty} y\right) = \lim_{x \to \infty}\frac{1}{x}\ln x = \lim_{x \to \infty}\frac{\ln x}{x}$$

By L'Hopital's Rule:

$$\ln\left(\lim_{x \to \infty} y\right) = \lim_{x \to \infty}\frac{\frac{1}{x}}{1}$$

$$\ln\left(\lim_{x \to \infty} y\right) = 0$$

Rewrite in exponential form:

$$\lim_{x \to \infty} y = e^0 = 1$$

Therefore, $\lim_{x \to \infty} x^{1/x} = 1$.

BRAIN TICKLERS Set # 18

1. Evaluate $\lim_{x \to 0} \frac{e^{3x} - 1}{\sin x}$.
2. Evaluate $\lim_{x \to \infty} \frac{\ln x}{2\sqrt{x}}$.
3. Evaluate $\lim_{x \to \frac{\pi}{4}} \frac{\tan\left(x - \frac{\pi}{4}\right)}{x - \frac{\pi}{4}}$.
4. Evaluate $\lim_{x \to 0^+} x^x$.

(Answers are on page 153.)

Curve Sketching

The derivatives of a function are very useful when graphing the curve of that function. Whether the graph of a function goes up or down; has high or low points; is flat or has either a peak or a point; or is curved, each of these characteristics all relate back to the derivatives of the function. Numerous tests can be performed that help to create a sketch of the curve of a function.

First Derivative Test

When given a continuous, differentiable function over an interval, finding the value of the first derivative of the function over that interval will indicate if the graph of the function is increasing, decreasing, or constant. The first derivative test can also be used to find *relative maximum* and *relative minimum* points of the function.

Increasing/Decreasing Intervals

Let $f(x)$ be defined on an interval, and let a and b denote points in that interval. One of the following three situations is occurring:

- $f(x)$ is increasing on the interval if $f(a) < f(b)$ whenever $a < b$.
- $f(x)$ is decreasing on the interval if $f(a) > f(b)$ whenever $a < b$.
- $f(x)$ is constant on the interval if $f(a) = f(b)$ for all points a and b.

1+2=3 MATH TALK!

To tell if a function is increasing, decreasing, or constant, the *y*-values over an interval need to be examined. For a function to be increasing over an interval (a, b) means that the *y*-value at a is smaller than the *y*-value at b. Graphically speaking, when moving from left to right over the interval, the graph is going up. For a function to be decreasing over an interval (a, b) means that the *y*-value at a is greater than the *y*-value at b. Graphically speaking, when moving from left to right over the interval, the graph is going down. For a function to be constant over an interval (a, b), the *y*-value at a is equal to the *y*-value at b. Graphically speaking, when moving from left to right over the interval, the graph is staying the same.

The first derivative test can be used to find intervals where the function is increasing, decreasing, or constant. The first derivative test is *painless* and involves five steps.

Step 1: Find the first derivative of the function with respect to x (or with respect to the independent variable).

Step 2: Find all critical values; these are the x-values where the derivative is equal to zero or is undefined.

Step 3: Place the critical values on a number line; this is known as making a sign diagram.

Step 4: Test each interval by substituting a test value into the first derivative to see if the derivative is either positive or negative.

Step 5: Determine the behavior of the function based on the sign of the first derivative within each interval.

The table below summarizes the behavior of $f(x)$ over an interval depending upon the sign of its first derivative.

First Derivative of $f(x)$ over an Interval (a, b)	Behavior of $f(x)$ over an Interval (a, b)
$f'(x) > 0$	The graph of $f(x)$ increases over (a, b).
$f'(x) < 0$	The graph of $f(x)$ decreases over (a, b).
$f'(x) = 0$	The graph of $f(x)$ is constant over (a, b).

Example 16:

Given the function $f(x) = x^3 - x$, find the intervals where $f(x)$ is increasing, decreasing, or constant.

Solution:

The first derivative test can be used to find intervals where the function is increasing, decreasing, or constant.

Step 1: Find the first derivative of the function with respect to x:

$$f'(x) = 3x^2 - 1$$

Step 2: Find all critical values; these are the x-values where the derivative is equal to zero or is undefined:

$$f'(x) = 3x^2 - 1 = 0$$
$$x = \pm\frac{1}{\sqrt{3}}$$

Step 3: Place the critical values on a number line; this is known as making a sign diagram:

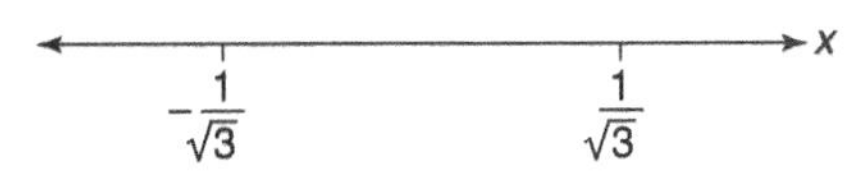

Step 4: Test each interval by substituting a test value into the first derivative to see if the derivative is either positive or negative:

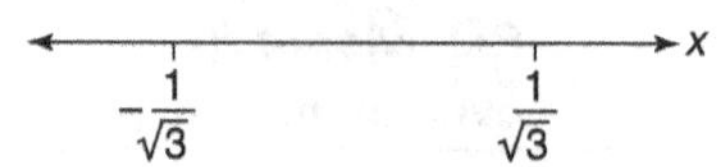

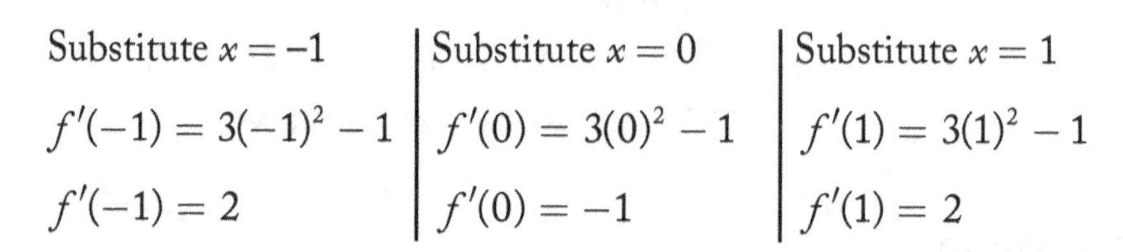

Substitute $x = -1$	Substitute $x = 0$	Substitute $x = 1$
$f'(-1) = 3(-1)^2 - 1$	$f'(0) = 3(0)^2 - 1$	$f'(1) = 3(1)^2 - 1$
$f'(-1) = 2$	$f'(0) = -1$	$f'(1) = 2$

Step 5: Determine the behavior of the function based on the sign of the first derivative within each interval.

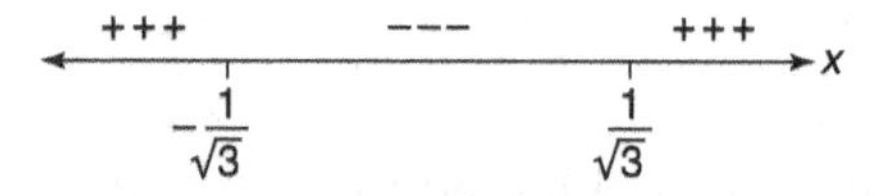

In the interval	In the interval	In the interval
$\left(-\infty, -\frac{1}{\sqrt{3}}\right)$,	$\left(-\frac{1}{\sqrt{3}}, \frac{1}{\sqrt{3}}\right)$,	$\left(-\frac{1}{\sqrt{3}}, \infty\right)$,
$f'(x) > 0$.	$f'(x) < 0$.	$f'(x) > 0$.
Therefore, $f(x)$ is increasing.	Therefore, $f(x)$ is decreasing.	Therefore, $f(x)$ is increasing.

This solution can be verified graphically.

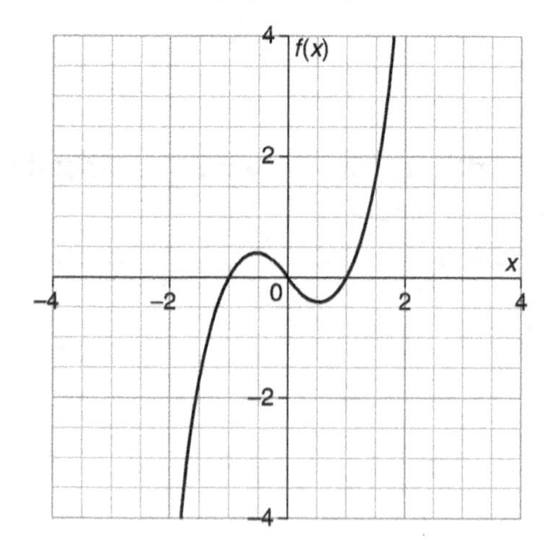

Starting on the left side of the graph and moving toward the right, the graph is going up until $x = -\frac{1}{\sqrt{3}}$. Between $x = -\frac{1}{\sqrt{3}}$ and $x = \frac{1}{\sqrt{3}}$, the graph is going down. The graph goes up again starting at $x = \frac{1}{\sqrt{3}}$. At each of the two critical values is a horizontal tangent line since the derivative is equal to zero at these values.

1+2=3 MATH TALK!

It is important to find all critical values of a function before creating a sign diagram. Critical values are where the derivative is either equal to zero or undefined. In other words, at these critical values, the tangent line is either horizontal or does not exist. If all the critical values are found, within each interval the slopes of all the tangent lines will have the same sign, either positive or negative.

Relative Extrema

A function is said to have a relative extremum or local extremum at an x-value if it is either a relative maximum or a relative minimum within an interval. It does not necessarily have to be the greatest or lowest value of the function over the entire interval. A function $f(x)$ is said to have a relative maximum at a if $f(a) \geq f(x)$ for all x in some open interval containing a. A function $f(x)$ is said to have a relative minimum at a if $f(a) \leq f(x)$ for all x in some open interval containing a.

The first derivative test can also be used to find the relative extrema of a function. Critical values that may be relative extrema are where the first derivative is either equal to zero or undefined. If a function is increasing to the left of a critical value and then decreasing to the right of that critical value, that point represents a relative maximum. If a function is decreasing to the left of a critical value and then increasing to the right of that critical value, that point represents a relative minimum.

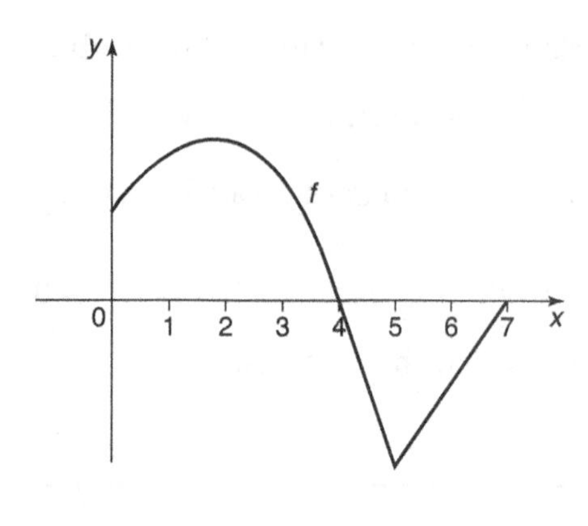

The figure to the left illustrates a relative maximum and a relative minimum.

At $x = 2$, there would be a horizontal tangent line, meaning $f'(2) = 0$. The function f is increasing to the left of $x = 2$ and decreasing to the right of $x = 2$. Therefore, $x = 2$ represents a relative maximum.

At $x = 5$, there is a corner, meaning $f'(5)$ is undefined. The function f is decreasing to the left of $x = 5$ and increasing to the right of $x = 5$. Therefore, $x = 5$ represents a relative minimum.

1+2=3 MATH TALK!

After finding the critical values of the first derivative and creating a sign diagram, if the derivative changes sign from positive to negative at a critical value, that *x*-value is a relative maximum. If the derivative changes sign from negative to positive at a critical value, that *x*-value is a relative minimum.

Example 17:

Let $f(x) = x^3 - 3x^2 - 9x + 10$. Find the intervals on which $f(x)$ is increasing and decreasing. Find all relative extrema.

Solution:

The first derivative test can be used to find intervals where the function is increasing and decreasing and to locate relative extrema.

Step 1: Find the first derivative of the function with respect to x:

$$f'(x) = 3x^2 - 6x - 9$$

Step 2: Find all critical values; these are the x-values where the derivative is equal to zero or undefined:

$$f'(x) = 3x^2 - 6x - 9 = 0$$

$$x^2 - 2x - 3 = 0$$

$$(x + 1)(x - 3) = 0$$

$$x = -1,\ x = 3$$

Step 3: Place the critical values on a number line; this is known as making a sign diagram.

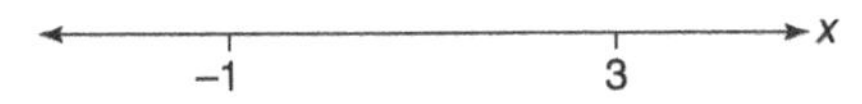

Step 4: Test each interval by substituting a test value into the first derivative to see if the derivative is either positive or negative.

Substitute $x = -2$	Substitute $x = 0$	Substitute $x = 4$
$f'(-2) = 3(-2)^2 - 6(-2) - 9$	$f'(0) = 3(0)^2 - 6(0) - 9$	$f'(4) = 3(4)^2 - 6(4) - 9$
$f'(-2) = 15$	$f'(0) = -9$	$f'(4) = 15$

Step 5: Determine the behavior of the function based on the sign of the first derivative within each interval.

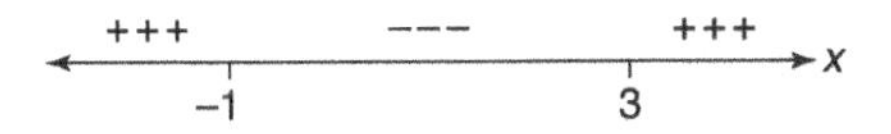

In the interval $(-\infty, -1)$, $f'(x) > 0$. Therefore, $f(x)$ is increasing.	In the interval $(-1, 3)$, $f'(x) < 0$. Therefore, $f(x)$ is decreasing.	In the interval $(3, \infty)$, $f'(x) > 0$. Therefore, $f(x)$ is increasing.

There is a relative maximum at $x = -1$ because the first derivative changes sign here from positive to negative.

There is a relative minimum at $x = 3$ because the first derivative changes sign here from negative to positive.

This solution can be verified graphically.

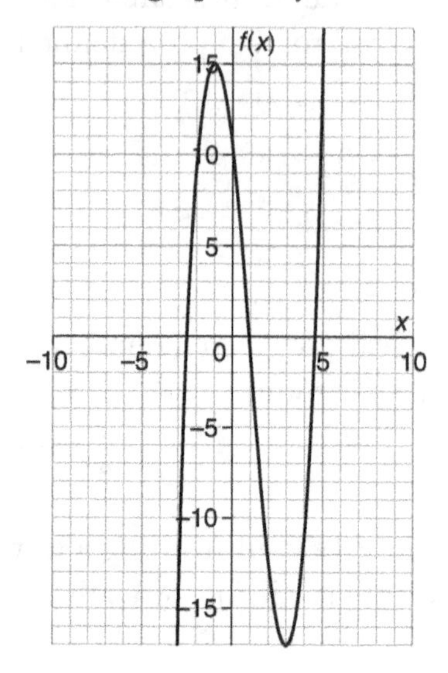

Example 18:

For the function $f(x) = x^2 + e^{-3x}$, describe the behavior of $f(x)$ at $x = 0$.

Solution:

The first derivative will show if the function is increasing or decreasing or will show if there are relative extrema at $x = 0$. Substituting $x = 0$ into the first derivative:

$$f'(x) = 2x - 3e^{-3x}$$
$$f'(0) = 2(0) - 3e^{-3(0)} = -3 < 0$$

Since the first derivative is negative at $x = 0$, $f(x)$ is decreasing at this point.

Example 19:

For the function f graphed below, identify all critical values and classify each as a relative maximum, a relative minimum, or neither.

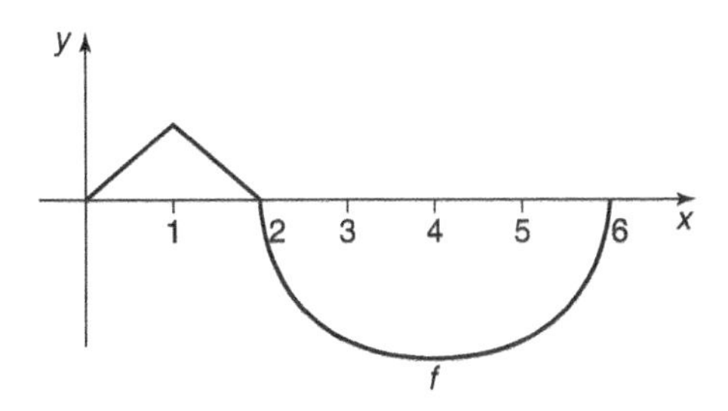

Solution:

Critical values are where the first derivative is equal to zero or undefined.

Critical Values	Reason	Classification	Reason
$x = 1$	There is a corner at $x = 1$, meaning the derivative is undefined.	Relative maximum	The function f is increasing to the left of $x = 1$ and decreasing to the right of $x = 1$.
$x = 2$	There is a cusp at $x = 2$, meaning the derivative is undefined.	Neither	The function f is decreasing to the left of $x = 2$ and decreasing to the right of $x = 2$. Since there is no change in behavior, the critical value is neither a relative maximum or a relative minimum.
$x = 4$	There is a horizontal tangent at $x = 4$, meaning the derivative is zero.	Relative minimum	The function f is decreasing to the left of $x = 4$ and increasing to the right of $x = 2$.

Example 20:

The derivative of function $f(x)$ is given by $f'(x) = \dfrac{x^2 - 9}{x}$ for all $x \neq 0$. Find all values of x for which the graph has relative extrema.

Solution:

The first derivative test can be used to find intervals where the function is increasing and decreasing and to locate relative extrema.

Step 1: You are given the first derivative of the function with respect to x:

$$f'(x) = \frac{x^2 - 9}{x}$$

Step 2: Find all critical values; these are the x-values where the derivative is equal to zero or undefined.

Derivative Equal to Zero	Derivative Undefined
Setting the numerator equal to zero,	Setting the denominator equal to zero,
$x^2 - 9 = 0$	$x = 0$
$(x + 3)(x - 3) = 0$	
$x = -3$, $x = 3$	

Step 3: Place the critical values on a number line; this is known as making a sign diagram.

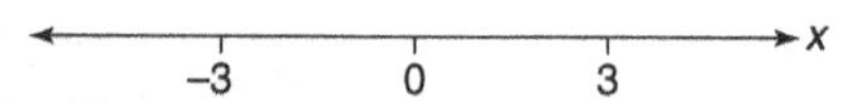

Step 4: Test each interval by substituting a test value into the first derivative to see if the derivative is either positive or negative.

x
–3 0 3

Substitute $x = -4$	Substitute $x = -1$	Substitute $x = 1$	Substitute $x = 4$
$f'(-4) = \dfrac{(-4)^2 - 9}{(-4)}$	$f'(-1) = \dfrac{(-1)^2 - 9}{(-1)}$	$f'(1) = \dfrac{(1)^2 - 9}{(1)}$	$f'(4) = \dfrac{(4)^2 - 9}{(4)}$
$f'(-4) = -1.75$	$f'(-1) = 8$	$f'(1) = -8$	$f'(1) = 1.75$

Step 5: Determine the behavior of the function based on the sign of the first derivative within each interval.

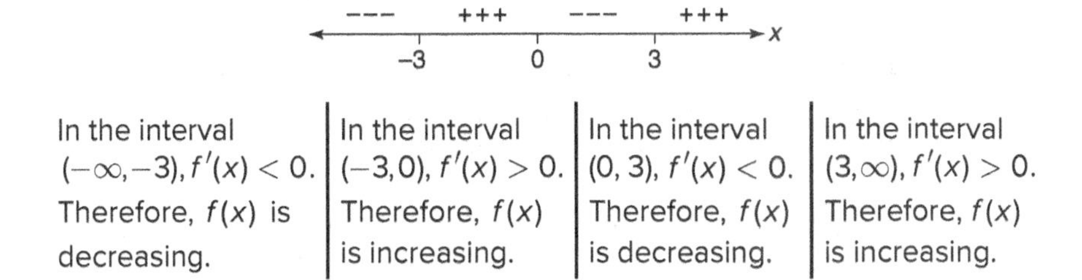

There is a relative minimum at $x = -3$ because the first derivative changes sign here from negative to positive.

There is a relative maximum at $x = 0$ because the first derivative changes sign here from positive to negative.

There is a relative minimum at $x = 3$ because the first derivative changes sign here from negative to positive.

Concavity

When given a continuous, differentiable function over an interval, finding the value of the second derivative of the function over an interval will show if the graph of the function is either concave up or concave down.

The concavity of a function refers to the curvature of the graph of the function. The graph can either be concave up, where the graph looks like it is curving up, or be concave down, where the graph looks like it is curving down.

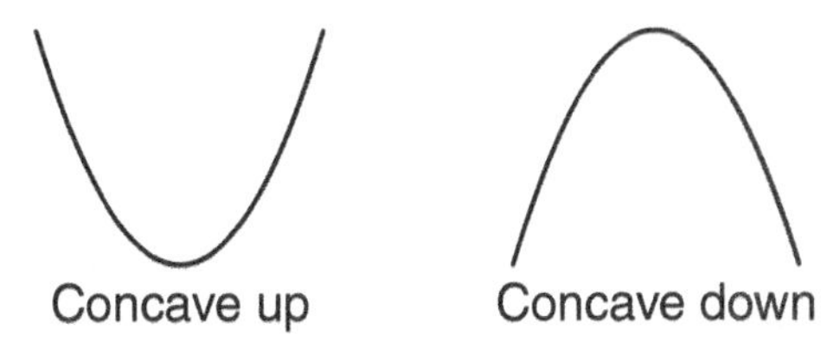

Let $f(x)$ be a differentiable function on an interval. The function $f(x)$ is concave up on the interval if $f''(x)$ is positive on the interval and is concave down on the interval if $f''(x)$ is negative on the interval.

A second derivative sign diagram can be used to find intervals where the function is concave up or concave down. Using the second derivative sign diagram is *painless* and involves five steps.

Step 1: Find the second derivative of the function with respect to x (or with respect to the independent variable).

Step 2: Find all critical values; these are the x-values where the second derivative is equal to zero or undefined.

Step 3: Create a sign diagram by placing the critical values on a number line.

Step 4: Test each interval by substituting a test value into the second derivative to see if the derivative is either positive or negative.

Step 5: Determine the behavior of the function based on the sign of the second derivative within each interval.

The table below summarizes the behavior of $f(x)$ over an interval depending upon the sign of its second derivative.

Second Derivative of $f(x)$ over an Interval (a, b)	Behavior of $f(x)$ over an Interval (a, b)
$f''(x) > 0$	The graph of $f(x)$ is concave up over (a, b).
$f''(x) < 0$	The graph of $f(x)$ is concave down over (a, b).

Example 21:

If $f(x) = x^3 - 3x^2 + 1$, find open intervals on which $f(x)$ is concave up and concave down.

Solution:

The second derivative can be used to find intervals where the function is concave up or is concave down.

Step 1: Find the second derivative of the function with respect to x:

$$f'(x) = 3x^2 - 6x$$
$$f''(x) = 6x - 6$$

Step 2: Find all critical values; these are the x-values where the second derivative is equal to zero or undefined:

$$f''(x) = 6x - 6 = 0$$
$$x = 1$$

Step 3: Create a sign diagram by placing the critical values on a number line.

Step 4: Test each interval by substituting a test value into the second derivative to see if the derivative is either positive or negative.

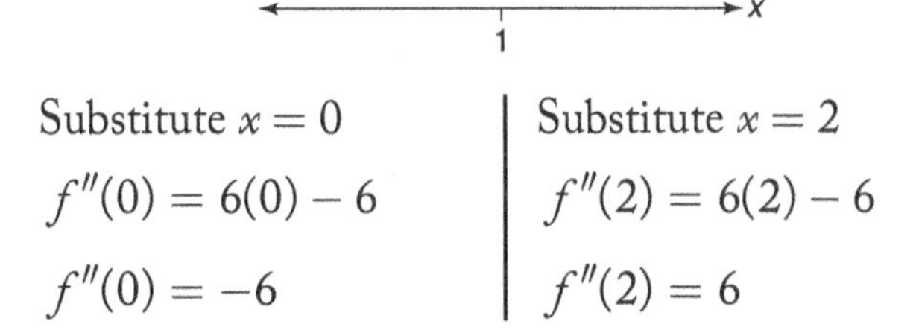

Substitute $x = 0$	Substitute $x = 2$
$f''(0) = 6(0) - 6$	$f''(2) = 6(2) - 6$
$f''(0) = -6$	$f''(2) = 6$

Step 5: Determine the behavior of the function based on the sign of the second derivative within each interval.

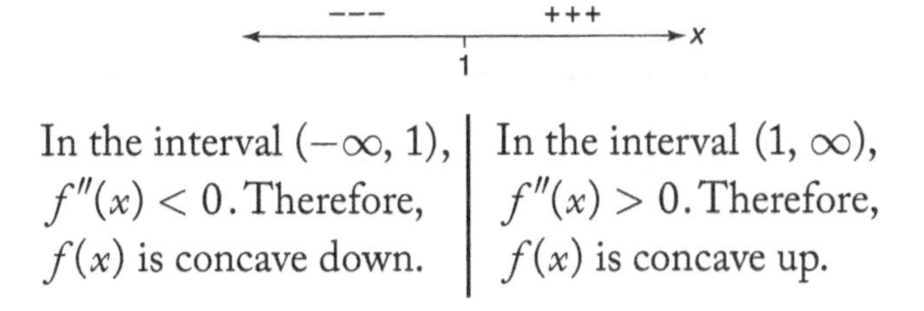

In the interval $(-\infty, 1)$, $f''(x) < 0$. Therefore, $f(x)$ is concave down.	In the interval $(1, \infty)$, $f''(x) > 0$. Therefore, $f(x)$ is concave up.

The solution can be verified graphically.

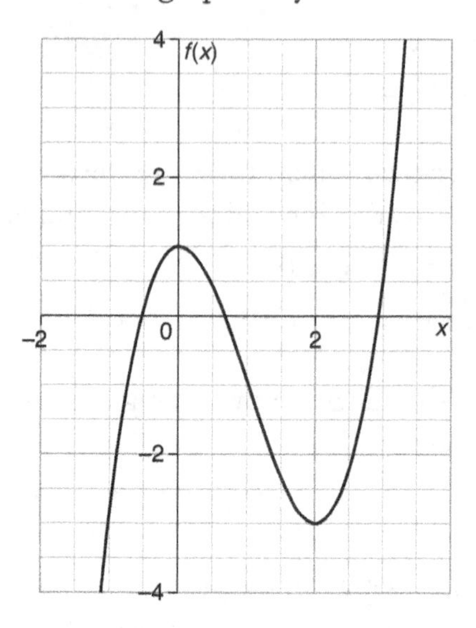

Point of Inflection

The point at which the graph of a function changes concavity is known as the point of inflection.

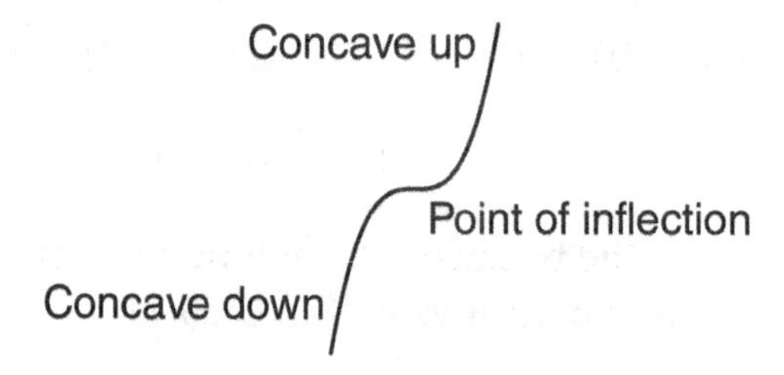

The second derivative can also be used to find the points of inflection of a function. After finding the critical values of the second derivative and creating a sign diagram, if the second derivative changes sign at a critical value, that x-value is a point of inflection.

Example 22:

Let $f(x) = x^4 + 2x^3 - 17x$. Find the open intervals on which $f(x)$ is concave up and concave down. Find all points of inflection.

Solution:

The second derivative can be used to find intervals where the function is concave up or concave down and points of inflection.

Step 1: Find the second derivative of the function with respect to x:

$$f'(x) = 4x^3 + 6x^2 - 17$$

$$f''(x) = 12x^2 + 12x$$

Step 2: Find all critical values; these are the x-values where the second derivative is equal to zero or undefined:

$$f''(x) = 12x^2 + 12x = 0$$

$$12x(x+1) = 0$$

$$x = 0,\ x = -1$$

Step 3: Create a sign diagram by placing the critical values on a number line.

Step 4: Test each interval by substituting a test value into the second derivative to see if the derivative is either positive or negative.

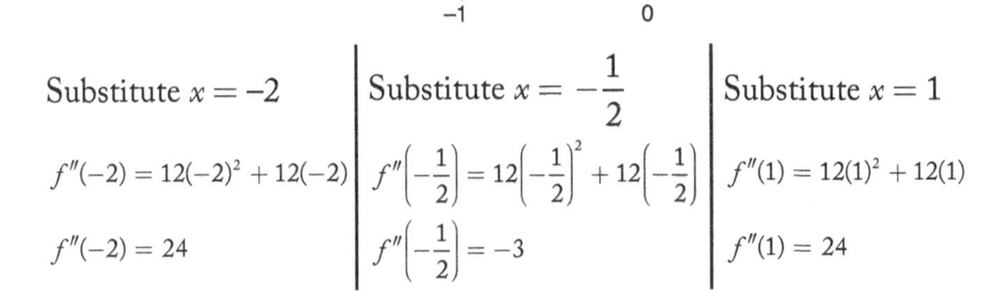

Substitute $x = -2$	Substitute $x = -\frac{1}{2}$	Substitute $x = 1$
$f''(-2) = 12(-2)^2 + 12(-2)$	$f''\left(-\frac{1}{2}\right) = 12\left(-\frac{1}{2}\right)^2 + 12\left(-\frac{1}{2}\right)$	$f''(1) = 12(1)^2 + 12(1)$
$f''(-2) = 24$	$f''\left(-\frac{1}{2}\right) = -3$	$f''(1) = 24$

Step 5: Determine the behavior of the function based on the sign of the second derivative within each interval.

In the interval $(-\infty, -1)$, $f''(x) > 0$. Therefore, $f(x)$ is concave up.	In the interval $(-1, 0)$, $f''(x) < 0$. Therefore, $f(x)$ is concave down.	In the interval $(0, \infty)$, $f''(x) > 0$. Therefore, $f(x)$ is concave up.

There are points of inflection at $x = -1$ and $x = 0$ since the second derivative changes sign around each of these critical values.

Example 23:

Using the first and second derivatives, sketch the graph of $f(x) = x^3 - 3x + 2$.

Solution:

The first derivative test will be conducted first to see where the function is increasing and decreasing and to identify relative extrema:

$$f'(x) = 3x^2 - 3 = 0$$

Critical values: $x = -1, x = 1$

Substitute $x = -2$	Substitute $x = 0$	Substitute $x = 2$
$f'(-2) = 3(-2)^2 - 3$	$f'(0) = 3(0)^2 - 3$	$f'(2) = 3(2)^2 - 3$
$f'(-2) = 9 > 0$	$f'(0) = -3 < 0$	$f'(2) = 9 > 0$
$f(x)$ is increasing in $(-\infty, -1)$	$f(x)$ is decreasing in $(-1, 1)$	$f(x)$ is increasing in $(1, \infty)$

There is a relative maximum at $x = -1$ because the derivative changes sign from positive to negative.

There is a relative minimum at $x = 1$ because the derivative changes sign from negative to positive.

The second derivative will be used to learn the concavity of the function and to locate points of inflection:

$$f''(x) = 6x = 0$$

Critical value: $x = 0$

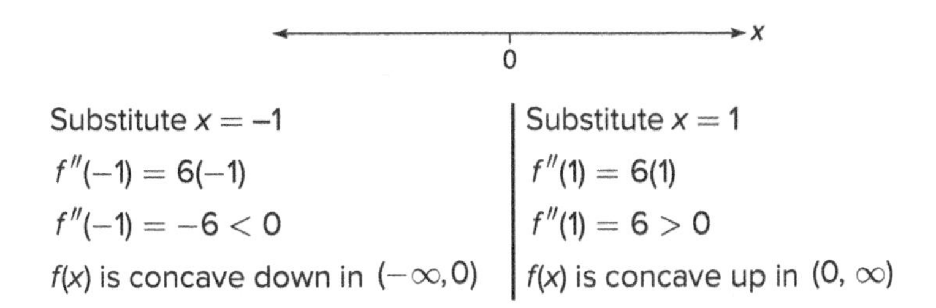

Substitute $x = -1$	Substitute $x = 1$
$f''(-1) = 6(-1)$	$f''(1) = 6(1)$
$f''(-1) = -6 < 0$	$f''(1) = 6 > 0$
$f(x)$ is concave down in $(-\infty, 0)$	$f(x)$ is concave up in $(0, \infty)$

There is a point of inflection at $x = 0$ because the second derivative changes sign from negative to positive.

This can be verified graphically.

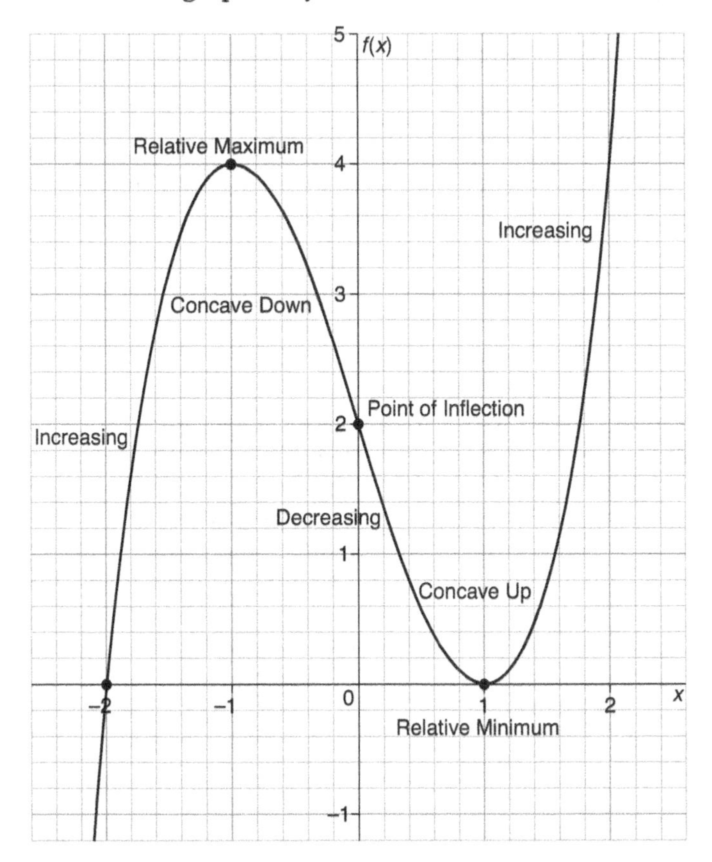

Example 24:

Find the coordinate(s) of the point(s) of inflection on the graph of $f(x) = (x + 2)\tan^{-1} x$.

Solution:

To find the points of inflection, use the second derivative:

$$f'(x) = (x+2)\left(\frac{1}{1+x^2}\right) + (\tan^{-1} x)(1) = (x+2)(1+x^2)^{-1} + \tan^{-1} x$$

$$f''(x) = (x+2) \bullet -1(1+x^2)^{-2} \bullet 2x + (1+x^2)^{-1}(1) + \left(\frac{1}{1+x^2}\right)$$

$$f''(x) = \frac{-2x(x+2)}{(1+x^2)^2} + \frac{1}{(1+x^2)} + \frac{1}{(1+x^2)} = \frac{-2x^2 - 4x + 1 + x^2 + 1 + x^2}{(1+x^2)^2}$$

$$f''(x) = \frac{-4x+2}{(1+x^2)^2}$$

The critical values are where $f''(x) = 0$ or is undefined.

Second Derivative Equal to Zero	Second Derivative Undefined
Set the numerator equal to zero	Set the denominator equal to zero
$-4x + 2 = 0$	$(1 + x^2)^2 = 0$
$x = \frac{1}{2}$	$1 + x^2 = 0$
	No solution

$$\longleftarrow \quad \frac{1}{2} \quad \longrightarrow x$$

Substitute values within each interval into the second derivative.

Substitute $x = 0$	Substitute $x = 1$
$f''(0) = \frac{-4(0)+2}{(1+(0)^2)^2}$	$f''(1) = \frac{-4(1)+2}{(1+(1)^2)^2}$
$f''(0) = \frac{2}{1} > 0$	$f''(1) = \frac{-2}{4} < 0$
$f(x)$ is concave up	$f(x)$ is concave down

There is one point of inflection at $x = \frac{1}{2}$ because the sign of $f''(x)$ changes. Its coordinates are $\left(\frac{1}{2}, f\left(\frac{1}{2}\right)\right) \approx \left(\frac{1}{2}, 1.159\right)$.

The concavity of a graph also determines if the graph of a tangent line is above the curve or below the curve. This is helpful when determining if the approximation of a function using the equation of a tangent line is either an overestimate or an underestimate. If the graph is concave up over an interval, the graph of the tangent line is drawn below the curve, making its tangent approximation an *underestimate.* If the graph is concave down over an interval, the graph of the tangent line is drawn above the curve, making its tangent approximation an *overestimate.*

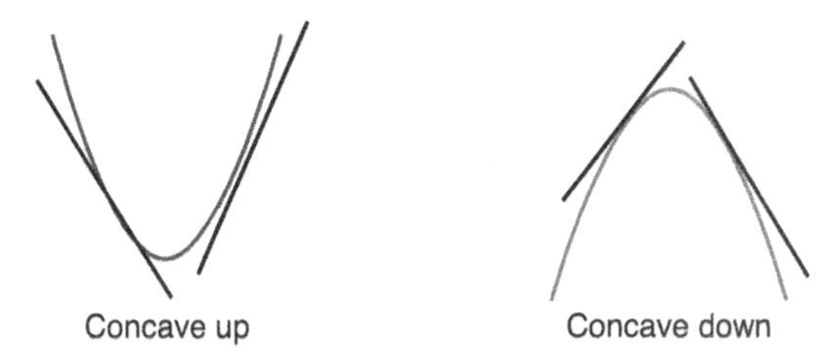

Example 25:

Let $h(x)$ be a function defined for all $x \neq 0$ such that $h(4) = -3$ and $h'(x) = \dfrac{x^2 - 2}{x}$, $x \neq 0$. Write an equation for the line tangent to the graph of $h(x)$ at $x = 4$, and explain why this line lies either above or below the graph of $h(x)$ for $x < 4$.

Solution:

The point of tangency is given as $(4, -3)$. To find the slope of the tangent line, evaluate $h'(4)$:

$$h'(4) = \frac{x^2 - 2}{x} = \frac{(4)^2 - 2}{(4)} = \frac{14}{4} = \frac{7}{2}$$

Write the equation of the tangent line at $(4, -3)$:

$$y + 3 = \frac{7}{2}(x - 4)$$

To determine if the tangent line lies above or below the graph of $h(x)$, evaluate $h''(4)$. This determines the concavity of the curve at $x = 4$:

$$h''(x) = \frac{x(2x) - (x^2 - 2)(1)}{x^2} = \frac{2x^2 - x^2 + 2}{x^2} = \frac{x^2 + 2}{x^2}$$

$$h''(4) = \frac{(4)^2 + 2}{(4)^2} = \frac{18}{16} > 0$$

The graph is concave up when $x = 4$. Therefore, the tangent line lies below the graph of $h(x)$ for $x < 4$. This means that any approximations from the tangent line would be underestimations for the curve.

BRAIN TICKLERS Set # 19

1. Find all open intervals for which $f(x) = x^3 - 3x^2$ is decreasing.
2. Find all values of x for which $f(x) = 3x^5 - 5x^3 - 12$ has a relative maximum.
3. Given the function $f(x) = 3x^5 - 20x^3$, find all values of x for which the graph of f is concave up.
4. Find the coordinates of the points of inflection for the graph of $y = 5x^4 - x^5$.

(Answers are on page 153.)

Graphical Representation of the Derivative

The derivatives of a function are useful when graphing the function. The graph of the derivative can also be used to discuss the behavior of the function. To graph the derivative, the y-axis changes from an axis for the values of $f(x)$ to the numerical values of $f'(x)$. Sign diagrams are also useful when graphing the derivative function.

CAUTION—Major Mistake Territory!

It can be confusing graphing the derivative function since the y-axis that usually represents the values of $f(x)$ now represents the values of $f'(x)$. A sign diagram can be useful when graphing either $f(x)$ or $f'(x)$. If the sign of the derivative is positive, the graph of $f'(x)$ will lie above the x-axis. If the sign of the derivative is negative, the graph of $f'(x)$ will lie below the x-axis. If the derivative is zero, that is where the graph of $f'(x)$ crosses the x-axis.

A table of functions, their derivative sign diagrams, and the graphs of their derivatives is on the facing page.

Function, $f(x)$	Sign Diagram	Derivative Function, $f'(x)$
$f(x) = x^2$	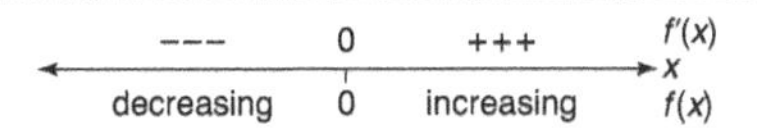	$f'(x) = 2x$
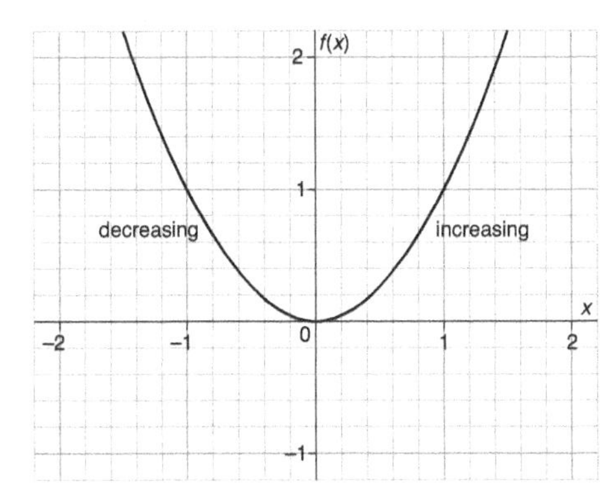		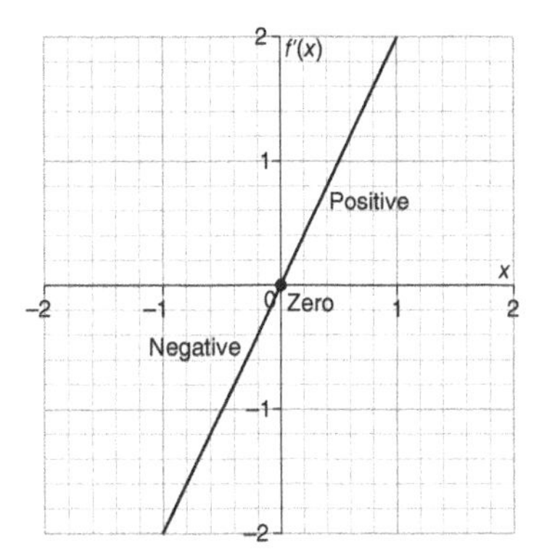
$f(x) = x^3$	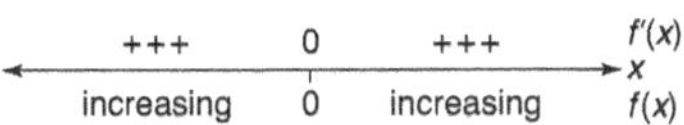	$f'(x) = 3x^2$
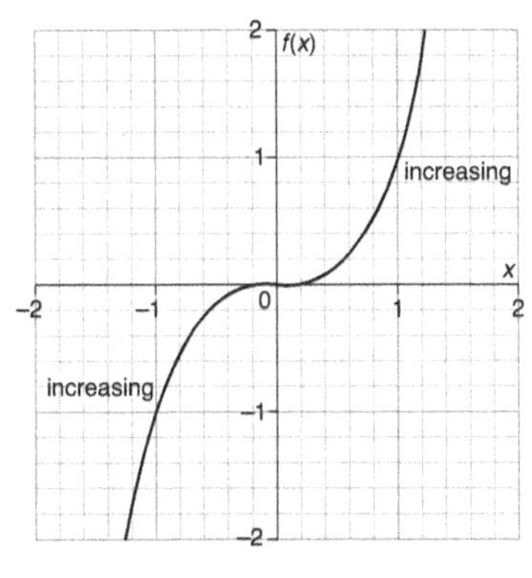		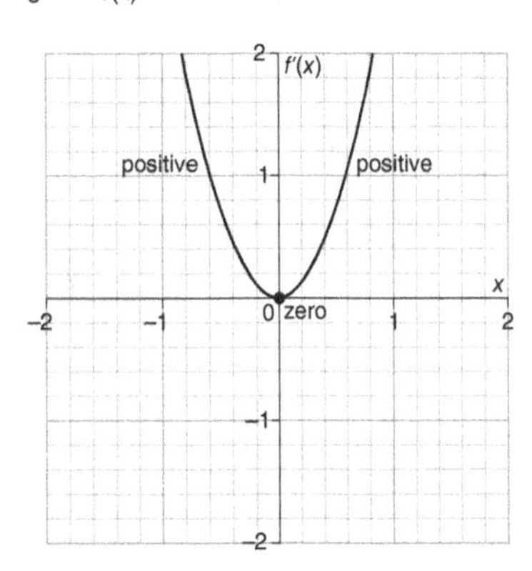
$f(x) = x^3 - 3x + 4$	$f'(x)$: +++ 0 −−− 0 +++ ; x: −1, 1 ; $f(x)$: increasing, decreasing, increasing	$f'(x) = 3x^2 - 3$
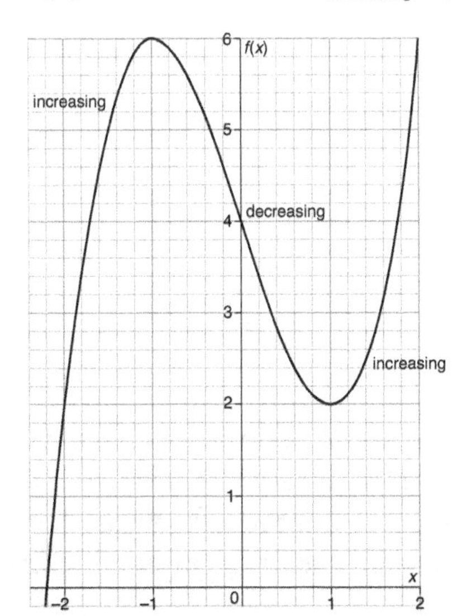		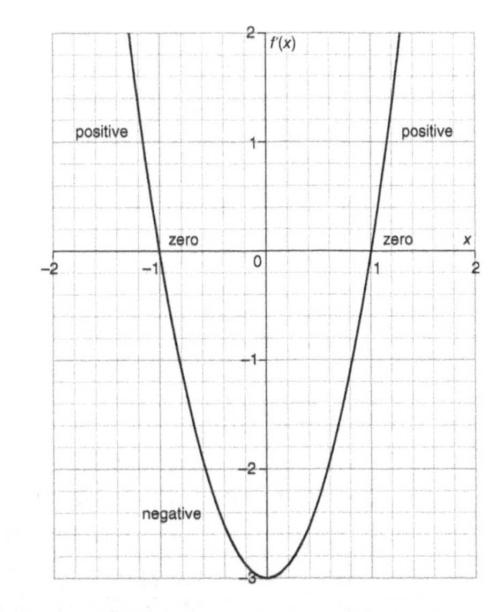

Example 26:

Below is a graph of $f'(x)$. Answer the following questions about $f(x)$ from the graph.

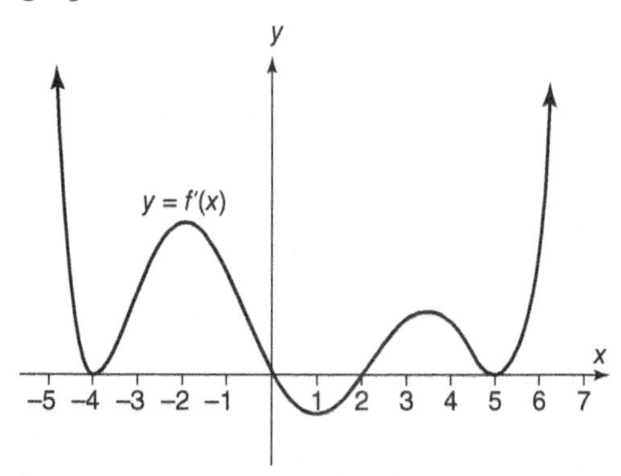

1. On what open interval(s) is f increasing?
2. At what x-values does the graph of $f(x)$ have a horizontal tangent line?
3. On the interval $-5 < x < 6$, find all x-values at which f attains a relative maximum.
4. On the interval $-5 < x < 6$, find all x-values at which f attains a relative minimum.

Solution:

1. A function is increasing where $f'(x) > 0$. This occurs when the graph of $f'(x)$ is above the x-axis in the intervals $(-\infty, -4) \cup (-4, 0) \cup (2, 5) \cup (5, \infty)$.
2. A function has a horizontal tangent line when $f'(x) = 0$. This occurs when the graph of $f'(x)$ intersects the x-axis at $x = -4, 0, 2, 5$.
3. A function attains a relative maximum at a critical value where the sign of the derivative changes from positive to negative. This occurs when $x = 0$ because coming in from the left toward 0, the graph of $f'(x)$ is above the x-axis and then the graph of $f'(x)$ is under the x-axis.
4. A function attains a relative minimum at a critical value where the sign of the derivative changes from negative to positive. This occurs when $x = 2$ because coming in from the left toward 2, the graph of $f'(x)$ is below the x-axis and then the graph of $f'(x)$ is above the x-axis.

Example 27:

The figure below depicts the graph of $f'(x)$ over the domain $-10 < x < 10$.

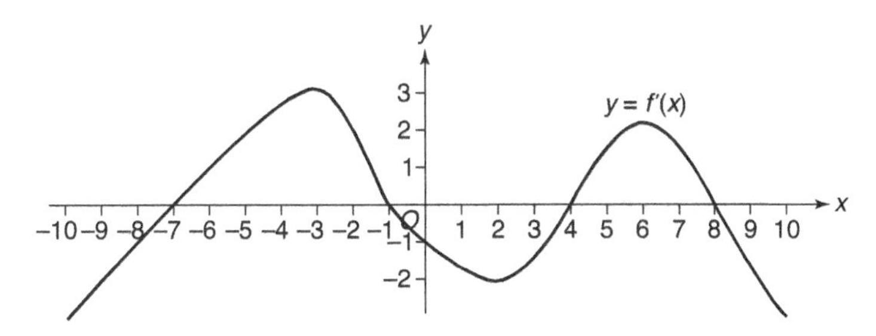

1. For what values of x does the graph of $f(x)$ have a horizontal tangent?
2. For what values of x in the interval $(-10, 10)$ does $f(x)$ have a relative maximum?
3. List the open intervals that include the values of x for which the graph of $f(x)$ is concave down.

Solution:

1. A function has a horizontal tangent when $f'(x) = 0$. This occurs when $x = -7, -1, 4, 8$.
2. A function attains a relative maximum at a critical value where the sign of the derivative changes from positive to negative or, in this case, when the graph changes from above the x-axis to below the x-axis. This occurs when $x = -1$ and $x = 8$.
3. A function is concave down when $f''(x) < 0$. When using the graph of $f'(x)$, $f''(x) < 0$ when the graph is decreasing. This occurs when x is in the intervals $(-3, 2) \cup (6, 10)$.

Second Derivative Test

The second derivative test is another way to determine relative extrema for a function. It states that if a function $f(x)$ is twice differentiable at $x = a$ and if $f'(a) = 0$, the following are true.

- If $f''(a) > 0$, then $f(x)$ has a relative minimum at $x = a$.
- If $f''(a) < 0$, $f(x)$ has a relative maximum at $x = a$.

In other words, if there is a horizontal tangent at $x = a$ and the curve is also concave up at this value, this point is a relative minimum. Similarly, if there is a horizontal tangent at $x = a$ and the curve is also concave down at this value, this point is a relative maximum. A relative minimum and a relative maximum are shown in the figures below.

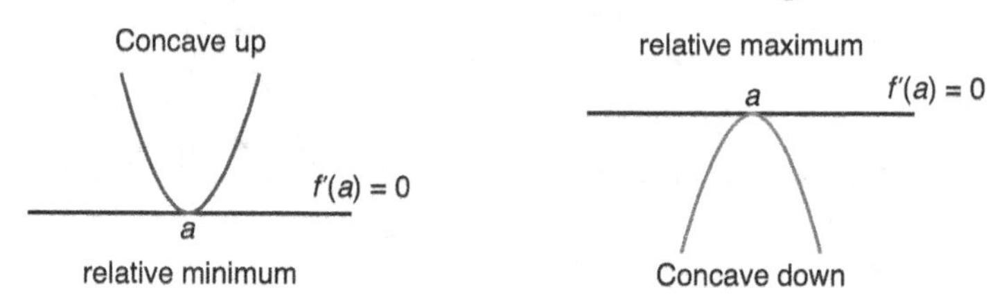

The second derivative test is *painless* and involves three steps.

Step 1: Find the critical values of the function, the x-values where the first derivative is equal to zero.

Step 2: Substitute the critical values into the second derivative of the function.

Step 3: Determine the behavior of the point based on the sign of the second derivative.

Example 28:

Locate and describe the relative extrema of $f(x) = x^4 - 2x^2$.

Solution:

One way to find the relative extrema is to use the second derivative test.

Step 1: Find the critical values of the function, the x-values where the first derivative is equal to zero:

$$f'(x) = 4x^3 - 4x = 0$$
$$4x(x^2 - 1) = 0$$
$$x = 0,\ x = -1,\ x = 1$$

Step 2: Substitute the critical values into the second derivative of the function:

$f''(x) = 12x^2 - 4$

$f''(-1) = 12(-1)^2 - 4 = 8$

$f''(0) = 12(0)^2 - 4 = -4$

$f''(1) = 12(1)^2 - 4 = 8$

Step 3: Determine the behavior of the point based on the sign of the second derivative.

$x = -1$ is a relative minimum because $f'(-1) = 0$ and $f''(-1) > 0$.

$x = 0$ is a relative maximum because $f'(0) = 0$ and $f''(0) < 0$.

$x = 1$ is a relative minimum because $f'(1) = 0$ and $f''(1) > 0$.

This can be verified graphically.

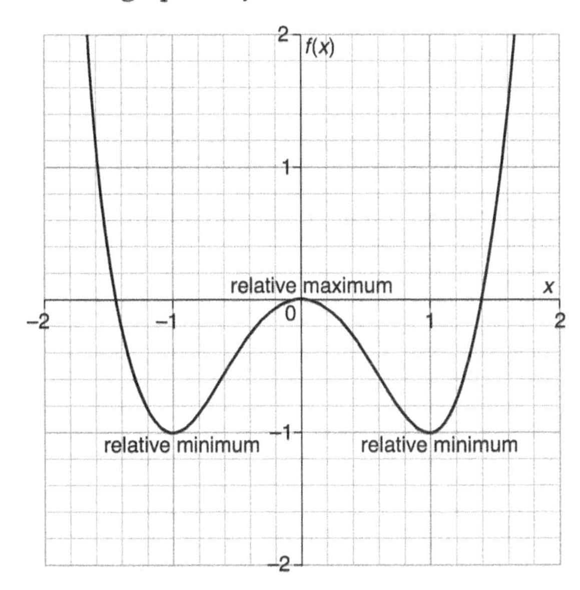

Absolute Maxima and Minima

A number that is either the maximum or minimum value of a function $f(x)$ is called an *absolute extreme value* or an *extreme value* of $f(x)$. If $f(a) \geq f(x)$ for all x in the domain of $f(x)$, then $f(a)$ is called the absolute maximum value. If $f(a) \leq f(x)$ for all x in the domain of $f(x)$, then $f(a)$ is called the absolute minimum value.

Absolute extrema can occur at critical values, where the first derivative equals zero or is undefined. They can also occur at the endpoints of a function.

Pictured below are different examples of absolute extrema and the x-values where they occur.

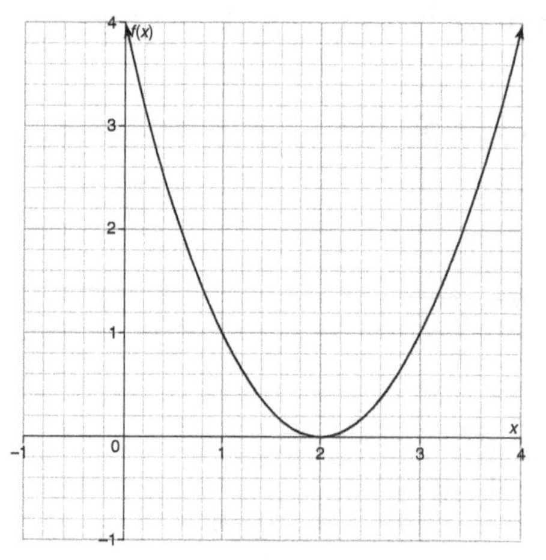

This graph has no absolute maximum since the end behavior of the graph continues to increase without bound.

There is an absolute minimum value of 0 at $x = 2$. This occurs at a critical value, which is also a relative minimum.

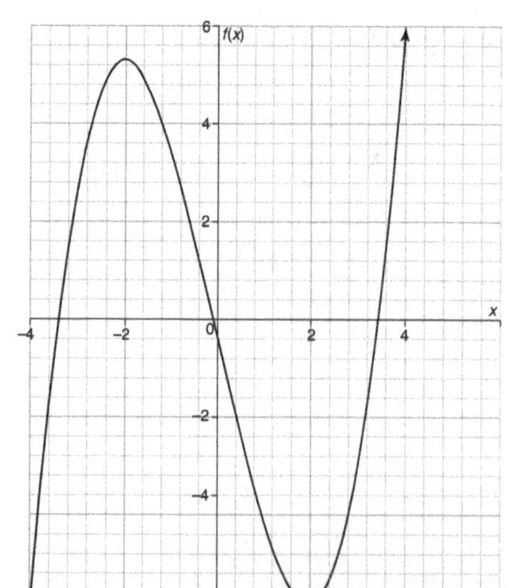

This graph has no absolute maximum since the end behavior of the graph continues to increase without bound.

This graph has no absolute minimum since the end behavior of the graph continues to decrease without bound.

There is a relative maximum at $x = -2$ and a relative minimum at $x = 2$.

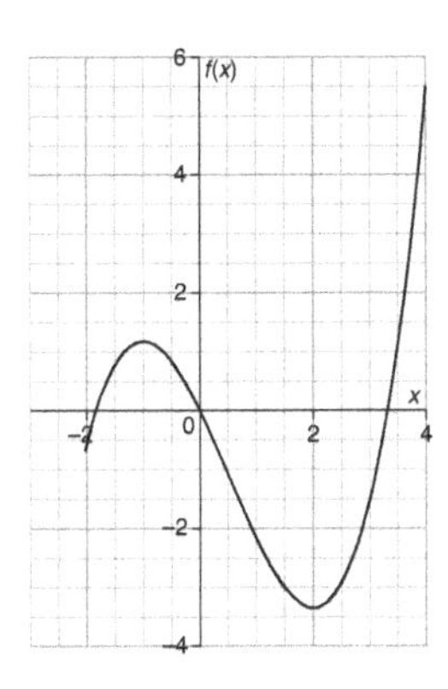

This graph has an absolute maximum value of 5.5 at $x = 4$. This occurs at an endpoint of the graph.

This graph has an absolute minimum value of $-\frac{10}{3} \approx -3.333$ at $x = 2$. This occurs at a critical value, which is also a relative minimum.

There is a relative maximum at $x = -1$.

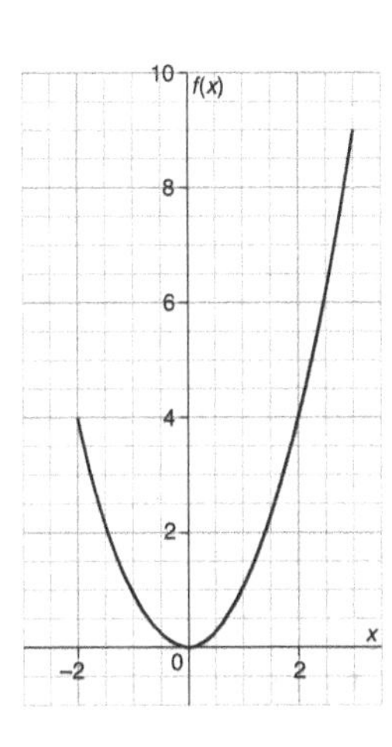

This graph has an absolute maximum value of 9 at $x = 3$. This occurs at an endpoint of the graph.

This graph has an absolute minimum value of 0 at $x = 0$. This occurs at a critical value, which is also a relative minimum.

Finding the absolute extrema of a function is *painless* and involves three steps.

Step 1: Find the critical values and endpoints of the function.

Step 2: Evaluate the function at each of these points.

Step 3: The largest function value is the absolute maximum, and the smallest function value is the absolute minimum.

Example 29:

Find the extreme maximum and minimum values for $f(x) = 2x^3 - 15x^2 + 36x$ over the interval $[1, 5]$.

Solution:

Step 1: Find the critical values and endpoints of the function.

The critical values for this function are where $f'(x) = 0$:

$f'(x) = 6x^2 - 30x + 36 = 0$

$(x - 2)(x - 3) = 0$

$x = 2, x = 3$

The endpoints of this function are $x = 1, x = 5$.

Step 2: Evaluate the function at each of these points.

x	$f(x) = 2x^3 - 15x^2 + 36x$
1	$f(1) = 2(1)^3 - 15(1)^2 + 36(1) = 23$
2	$f(2) = 2(2)^3 - 15(2)^2 + 36(2) = 28$
3	$f(3) = 2(3)^3 - 15(3)^2 + 36(3) = 27$
5	$f(5) = 2(5)^3 - 15(5)^2 + 36(5) = 55$

Step 3: The largest function value is the absolute maximum, and the smallest function value is the absolute minimum.

The absolute maximum is 55 when $x = 5$, and the absolute minimum is 23 when $x = 1$.

CAUTION—Major Mistake Territory!

When asked for the absolute extrema, keep in mind this is referring to the *y*-values of the function. You can also include the *x*-value, but this does not represent the absolute value.

Extreme Value Theorem

The *Extreme Value Theorem* states that if a function $f(x)$ is continuous on a closed interval $[a, b]$, then $f(x)$ has both a maximum and a minimum value on $[a, b]$.

Optimization

Optimization is an application of the absolute extrema of a function. It is applied to a real-life situation that can be modeled as a function. By applying the second derivative test, an absolute maximum or absolute minimum can be found for the scenario, thereby optimizing the function. The steps to optimizing a function are *painless*.

Step 1: Create a diagram that models the situation.

Step 2: Define your variables, and determine a domain.

Step 3: Create a function that represents what is to be optimized.

Step 4: Identify any absolute extrema by performing the second derivative test or creating a table of values.

Example 30:

Joseph has decided to build a rectangular garden outside of his home. He has 400 feet of fencing. What is the area of the largest rectangular garden he can enclose if he uses the building as one side of the garden?

Solution:

Step 1: Create a diagram that models the situation.

The first sentence describes a rectangular model. The second sentence gives 400 feet of fencing, which refers to the perimeter. The last sentence explains that only three sides of the rectangular model will be made using the fencing. Based upon this information, the following variables and diagram can be determined.

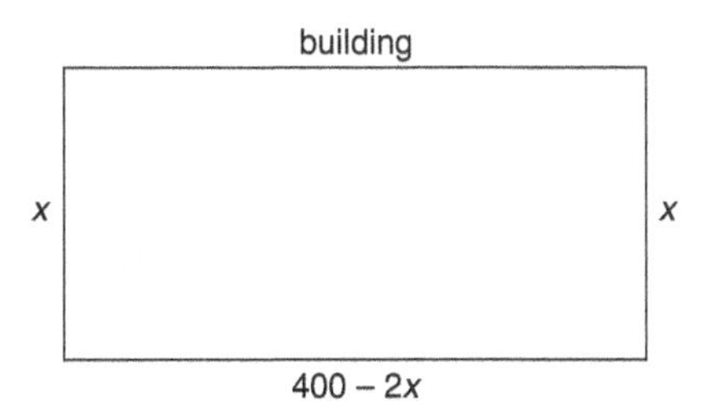

Step 2: Define your variables, and determine a domain.

Let $x =$ the width of the rectangular garden

Let $400 - 2x =$ the length of the rectangular garden

To determine a domain, each side of the rectangle must be greater than zero and the perimeter must be less than or equal to 400. Therefore, $400 - 2x > 0$. Solving for x yields the domain: $0 < x < 200$.

Step 3: Create a function that represents what is to be optimized.

The last sentence asks for the largest area. So, an equation must be created that models the area of the rectangle. Since the area of a rectangle is equal to the length multiplied by the width:

$$A(x) = (400 - 2x)(x)$$

$$A(x) = 400x - 2x^2$$

Step 4: Identify any absolute extrema by performing the second derivative test or creating a table of values.

To perform the second derivative test to find the absolute maximum, first find all critical values:

$$A'(x) = 400 - 4x = 0$$

$$x = 100$$

Since this is the only critical value and since the endpoints are not included in the domain, this is the only x-value that can yield an absolute extremum. To see if it will yield an absolute maximum value, substitute $x = 100$ into the second derivative:

$$A''(x) = -4$$

$$A''(100) = -4 < 0$$

Since $A'(100) = 0$ and $A''(100) < 0$, then $x = 100$ yields an absolute maximum value of:

$$A(100) = 400(100) - 2(100)^2 = 20,000$$

Therefore, the largest area of the rectangular garden is 20,000 square feet when the width is 100 feet and the length is 200 feet.

Example 31:

Kristen is going to make an open box from an 8-inch by 15-inch piece of cardboard by cutting out squares of equal size from the four corners and bending up the sides. What size should the squares be to obtain a box with the largest possible volume?

Solution:

Step 1: Create a diagram that models the situation.

The first sentence describes a rectangular prism model that is constructed from a rectangle. Based upon this information, the following variables and diagrams can be determined.

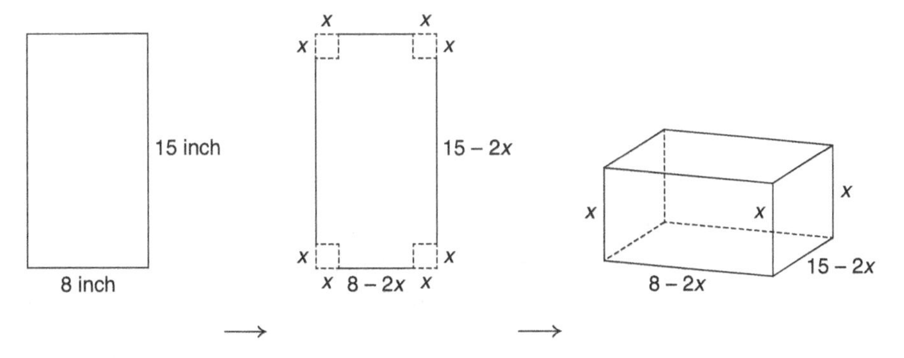

Note: the corners are folded up to make the height of the box, which is why the height is x.

Step 2: Define your variables, and determine a domain.

Let x = the height of the box

Let $8 - 2x$ = the length of the box

Let $15 - 2x$ = the width of the box

To determine a domain, each side of the box must be greater than zero, the length must be less than 8 inches, and the width must be less than 15 inches. Therefore, $8 - 2x > 0$. Solving for x yields the domain: $0 < x < 4$.

Step 3: Create a function that represents what is to be optimized.

The last sentence asks for the largest volume. So, an equation must be created that models the volume of the box. Since the volume of a rectangular prism is equal to the length multiplied by the width multiplied by the height:

$$V(x) = (8 - 2x)(15 - 2x)(x)$$

$$V(x) = 4x^3 - 46x^2 + 120x$$

Step 4: Identify any absolute extrema by performing the second derivative test or creating a table of values.

To perform the second derivative test to find the absolute maximum, first find all critical values:

$$V'(x) = 12x^2 - 92x + 120 = 0$$

$$3x^2 - 23x + 30 = 0$$

$$(3x - 5)(x - 6) = 0$$

$$x = \frac{5}{3},\ x = 6$$

The only critical value in the domain is $x = \frac{5}{3}$, so reject $x = 6$. Since there is only one critical value and since the endpoints are not included in the domain, this is the only x-value that can yield an absolute maximum. To see if it will yield an absolute maximum value, substitute $x = \frac{5}{3}$ into the second derivative:

$$V''(x) = 24x - 92$$

$$V''\left(\frac{5}{3}\right) = 24\left(\frac{5}{3}\right) - 92 = -52 < 0$$

Since $V'\left(\frac{5}{3}\right) = 0$ and $V''\left(\frac{5}{3}\right) < 0$, then $x = \frac{5}{3}$ yields an absolute maximum value of $V\left(\frac{5}{3}\right) = 4\left(\frac{5}{3}\right)^3 - 46\left(\frac{5}{3}\right)^2 + 120\left(\frac{5}{3}\right) = \frac{2450}{27}$ cubic inches. Therefore, the dimensions of the squares should be $\frac{5}{3}$ inches by $\frac{5}{3}$ inches to obtain a box with the largest possible volume.

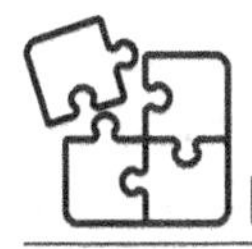

BRAIN TICKLERS Set # 20

1. The figure below shows the graph of $f'(x)$, which is the derivative of a function $f(x)$. For what values of x, $-3 < x < 3$, does $f(x)$ have a relative maximum, relative minimum, and point of inflection?

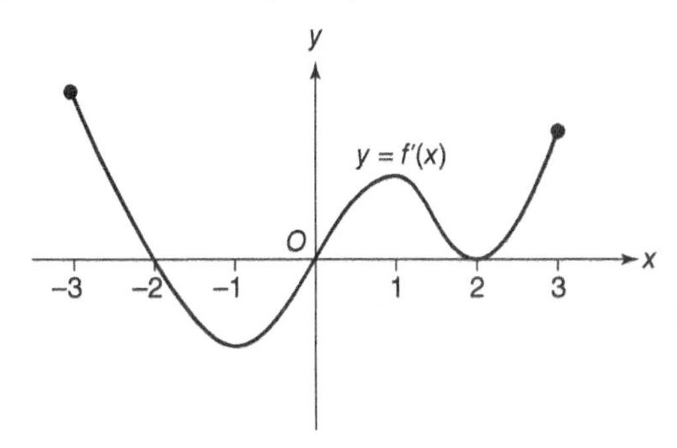

2. Let $f(x)$ be continuous and differentiable on the closed interval $[a, b]$. If $f(x)$ has a relative maximum at c and $a < c < b$, what can be said about $f'(c)$ and $f''(c)$?

3. If $f(x) = \frac{1}{3}x^3 - 4x^2 + 12x - 10$ and the domain is the set of all x such that $0 \le x \le 10$, find the absolute maximum value of the function $f(x)$.

4. An open box is to be made from a 16-inch by 30-inch piece of cardboard by cutting out squares of equal size from the four corners and bending up the sides. What size should the squares be to obtain a box with the largest volume?

(Answers are on page 153.)

Rectilinear Motion

The derivative of a function $f(x)$ at a real number c has an important interpretation as the instantaneous rate of change. An application of this interpretation in physics is the motion of a particle on a line, which is referred to as rectilinear motion.

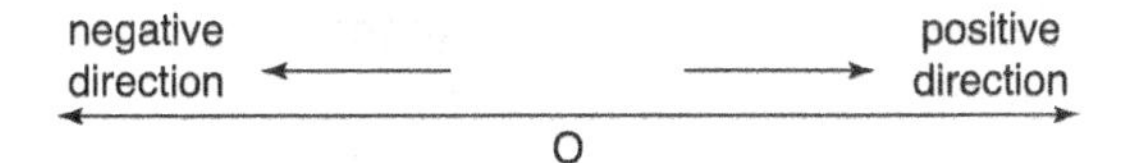

If the position of a particle, P, is to the left of position O, that position is a negative value. If the position of P is to the right of O, that position is a positive value.

Typically, the function that notates the position of particle P is referred to as $s(t)$. The velocity of P at time t is the derivative of $s(t)$ and is notated as $v(t)$. The acceleration of P at time t is the derivative of $v(t)$, which is the second derivative of $s(t)$, and is notated as $a(t)$.

Example 32:

Given the particle P, let s meters be the directed distance of the particle from O at t seconds defined by the equation $s(t) = t^2 + 2t - 3$, $t \geq 0$. Fill in the conclusion column below regarding the position of particle P at the selected times.

t (seconds)	*s(t)* (meters)	Conclusion
0	−3	
1	0	
2	5	
3	12	

Solution:

t (seconds)	*s(t)* (meters)	Conclusion
0	−3	Initially, P is 3 meters to the left of O.
1	0	At $t = 1$ second, P is at O.
2	5	At $t = 2$ seconds, P is 5 meters to the right of O.
3	12	At $t = 3$ seconds, P is 12 meters to the right of O.

Total Distance Versus Displacement

Distance is a scalar quantity representing the interval between two points; it is the magnitude of the interval. Distance depends on the whole path traveled between two points, which is not necessarily the shortest path. *Displacement* is a vector quantity. It is defined as the distance between the initial point and the final point of an object. Displacement must be the shortest interval connecting the initial and final points.

1+2=3 MATH TALK!

In other words, distance does not consider direction. Think of distance like an absolute value. Even if the particle has a negative s-value, the distance will add in that amount as a positive value. In contrast, displacement just considers the line segment from the initial point to the final point and does not take into account changes in direction. With displacement, you must be careful. The displacement might appear to be a smaller number than the distance, but that does not mean less distance was traveled.

Example 33:

Diane walks from point A to point B to point C. What is the distance she traveled? What is the displacement?

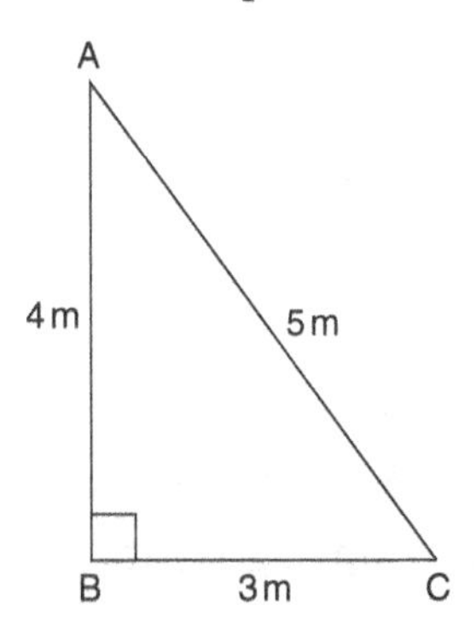

Solution:

$$\text{Distance} = 4\text{ m} + 3\text{ m} = 7\text{ m}$$

$$\text{Displacement} = 5\text{ m}$$

Example 34:

Chris leaves home and travels 20 miles to work. When he arrives, he realizes he left his briefcase at home. He drives back to get it and then returns to work. What is the distance he traveled? What is his displacement?

Solution:

Distance $= 20$ miles $+ 20$ miles $+ 20$ miles $= 60$ miles

Displacement $= 20$ miles $- 20$ miles $+ 20$ miles $= 20$ miles or

Displacement $= 20$ miles $- 0$ miles $= 20$ miles

Below is a summary of the different rectilinear motion functions and how they can be used to describe a particle's motion.

Calculus and Rectilinear Motion Summary

$s(t) =$ position function

$s'(t) = v(t) =$ velocity function

$s''(t) = v'(t) = a(t) =$ acceleration function

If $s(t) = 0$, the particle is at O.

If $s(t) > 0$, the particle is to the right of O.

If $s(t) < 0$, the particle is to the left of O.

If $v(t) = 0$, the particle is not moving.

If $v(t) = 0$ and the velocity changes sign to the left and right of this t-value, the particle has changed direction.

If $v(t) > 0$, the particle is moving right.

If $v(t) < 0$, the particle is moving left.

If $a(t) = 0$, the velocity is constant.

If $a(t) > 0$, the velocity is increasing.

If $a(t) < 0$, the velocity is decreasing.

Example 35:

A particle is moving along a horizontal line according to the equation $s(t) = t^3 - 12t^2 + 36t - 24,\ t \geq 0$.

1. Determine the interval of time when the particle is moving to the right and when it is moving to the left.
2. Determine the instant the particle reverses direction.

3. Find the total distance and displacement of the particle for the first 8 seconds.

Solution:

1. To find when a particle is moving to the right or left, the velocity equation must be found by taking the derivative of $s(t)$:

   $$v(t) = s'(t) = 3t^2 - 24t + 36$$

   Set $v(t) = 0$ and solve for t in order to find the critical values, where the particle has stopped moving:

   $$v(t) = 3t^2 - 24t + 36 = 0$$
   $$t^2 - 8t + 12 = 0$$
   $$(t - 6)(t - 2) = 0$$
   $$t = 6, t = 2$$

   To find the velocity to the left and right of each critical value, a sign diagram can be used.

   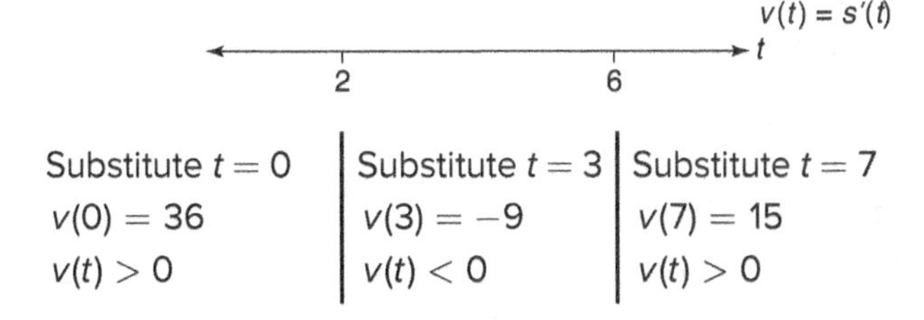

   The particle is moving to the right for the intervals of time $[0, 2)$ and $(6, \infty)$ because $v(t) > 0$ in those intervals. The particle is moving to the left in the time interval of $(2, 6)$ because $v(t) < 0$ in that interval.

2. The particle reverses direction at $t = 2$ and $t = 6$ because at both of those times, the velocity is equal to zero and the velocity changes sign at each of those t-values.

   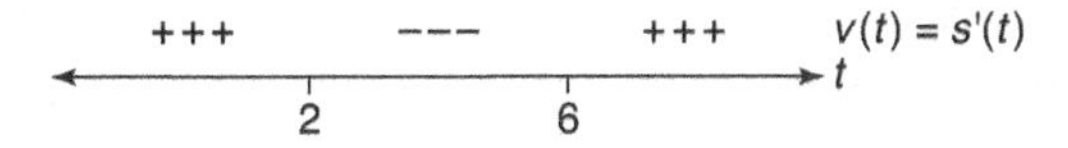

3. The most organized way to find the total distance would be to create a table of values for t and $s(t)$. The t-values that need to be included are the initial and final times and the times where $v(t) = 0$.

t	$s(t)$
0	−24
2	8
6	−24
8	8

For total distance, consider the position of the particle so that all distances are included. Between $t = 0$ and $t = 2$, the particle travels 32 units. Between $t = 2$ and $t = 6$, the particle travels 32 units. Between $t = 6$ and $t = 8$, the particle travels 32 units. In total, the distance equals:

$$32 + 32 + 32 = 96 \text{ units}$$

For displacement, consider only the initial and final positions. The displacement equals:

$$8 - (-24) = 32 \text{ units}$$

The speed of a particle at any time is defined as the absolute value of the instantaneous velocity or the magnitude of the velocity, $|v|$. Speed is always a nonnegative number. It is a scalar quantity and indicates only how fast the particle is moving. In contrast, velocity is a vector quantity. So, it also tells the direction of movement.

To determine if a particle is speeding up or slowing down at a specific time, the velocity and the acceleration of the particle need to be determined. For the particle to be speeding up, $v > 0$ and $a > 0$ or $v < 0$ and $a < 0$. In other words, the velocity and acceleration must either both be positive or both be negative. For the particle to be slowing down, $v > 0$ and $a < 0$ or $v < 0$ and $a > 0$. In other words, both the velocity and acceleration must have opposite signs; one is positive and the other is negative.

Example 36:

Using the position equation from Example 35, $s(t) = t^3 - 12t^2 + 36t - 24$, determine when the particle is speeding up or slowing down for $0 < t < 8$. Then determine the speed when $t = 3$.

Solution:

Since the signs of the velocity and acceleration equations need to be determined, create a sign diagram. Find the critical values for each equation by setting the velocity and acceleration equations equal to 0 and solving for t.

Set the velocity equal to zero.

$v(t) = s'(t) = 3t^2 - 24t + 36$

$3t^2 - 24t + 36 = 0$

$t^2 - 8t + 12 = 0$

$(t - 2)(t - 6) = 0$

$t = 2, t = 6$

Set the acceleration equal to zero.

$a(t) = v'(t) = 6t - 24$

$6t - 24 = 0$

$6t = 24$

$t = 4$

Create a sign diagram by substituting t-values from each of the three intervals into $v(t)$.

Create a sign diagram by substituting t-values from each of the two intervals into $a(t)$.

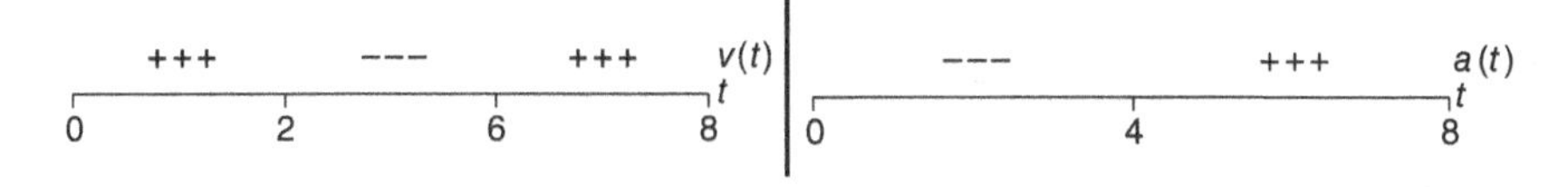

For $0 < t < 2$, the particle is slowing down since $v(t) > 0$ and $a(t) < 0$.

For $2 < t < 4$, the particle is speeding up since $v(t) < 0$ and $a(t) < 0$.

For $4 < t < 6$, the particle is slowing down since $v(t) < 0$ and $a(t) > 0$.

For $6 < t < 8$, the particle is speeding up since $v(t) > 0$ and $a(t) > 0$.

The speed when $t = 3$ is found by finding the magnitude of $v(3)$:

$$|v(3)| = |3(3)^2 - 24(3) + 36| = |-9| = 9$$

Example 37:

Use the graphs below to describe the characteristics of each curve at $t = t_0$ and then describe how each characteristic determines the behavior of the particle at $t = t_0$.

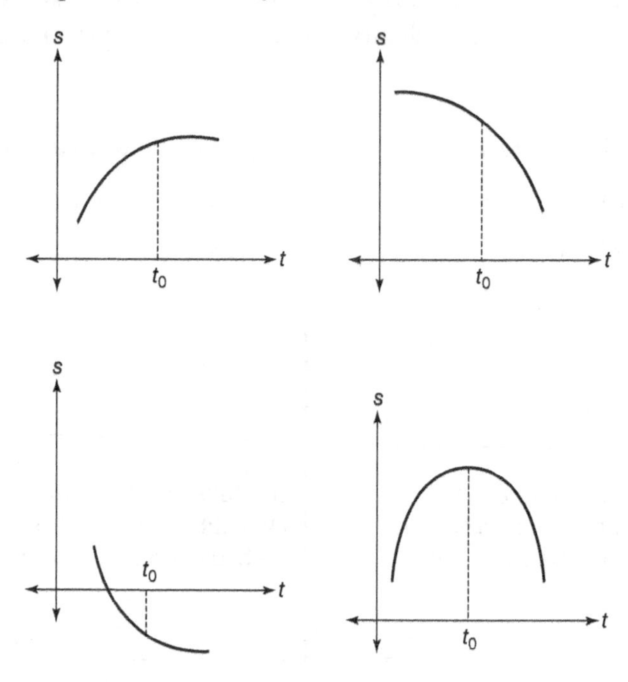

Solution:

Position vs. Time Curve	Characteristics of Curve at $t = t_0$	Behavior of Particle at $t = t_0$
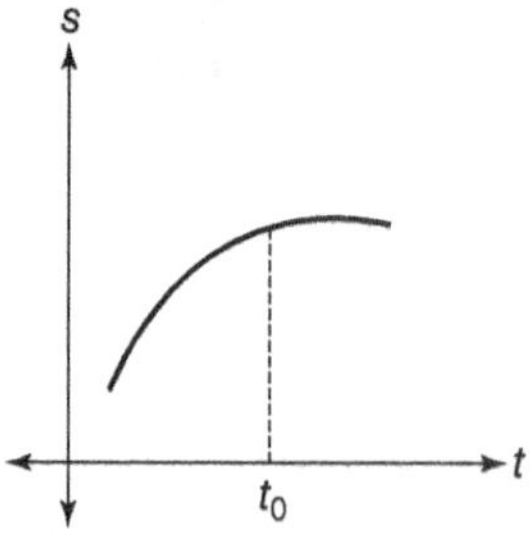	$s(t_0) > 0$ (graph is above t-axis)	The particle is to the right of O at $t = t_0$.
	$s'(t_0) > 0$ (slope of tangent at t_0 is positive)	The particle is moving to the right at $t = t_0$.
	$s''(t_0) < 0$ (graph is concave down)	The particle's velocity is decreasing at $t = t_0$.
		The particle is slowing down.

Position vs. Time Curve	Characteristics of Curve at $t = t_0$	Behavior of Particle at $t = t_0$
	$s(t_0) > 0$ (graph is above t-axis)	The particle is to the right of O at $t = t_0$.
	$s'(t_0) < 0$ (slope of tangent at t_0 is negative)	The particle is moving to the left at $t = t_0$.
	$s''(t_0) < 0$ (graph is concave down)	The particle's velocity is decreasing at $t = t_0$.
		The particle is speeding up.
	$s(t_0) < 0$ (graph is below t-axis)	The particle is to the left of O at $t = t_0$.
	$s'(t_0) < 0$ (slope of tangent at t_0 is negative)	The particle is moving to the left at $t = t_0$.
	$s''(t_0) > 0$ (graph is concave up)	The particle's velocity is increasing at $t = t_0$.
		The particle is slowing down.
	$s(t_0) > 0$ (graph is above t-axis)	The particle is to the right of O at $t = t_0$.
	$s'(t_0) = 0$ (slope of tangent at t_0 is zero)	The particle is not moving at $t = t_0$.
	$s''(t_0) < 0$ (graph is concave down)	The particle's velocity is decreasing at $t = t_0$.

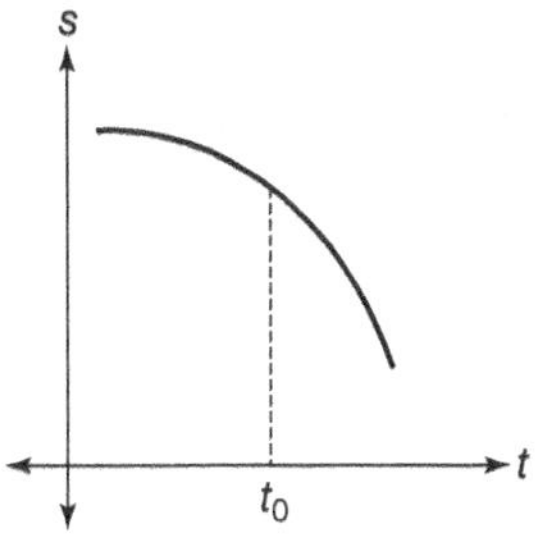

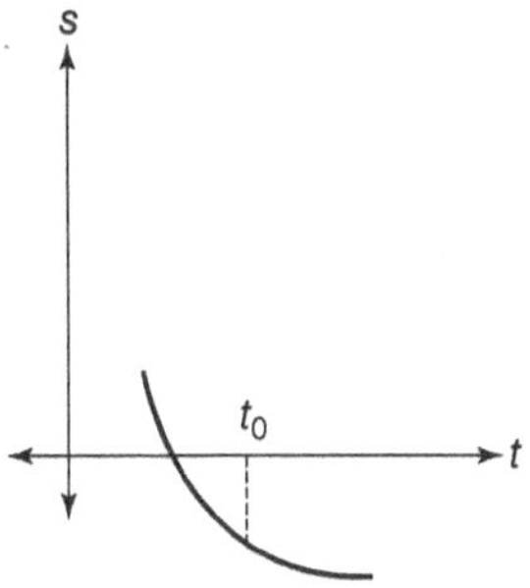

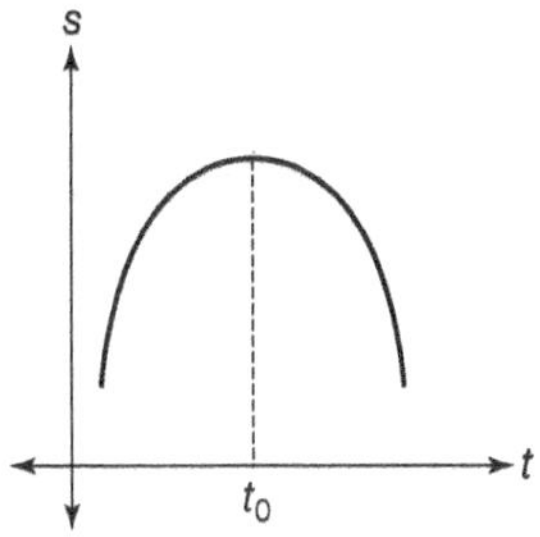

BRAIN TICKLERS Set # 21

1. A particle moves along a line according to the position equation $s(t) = 2t^3 - 9t^2 + 12t - 4$, where $t \geq 0$. Find all t for which the particle is moving to the right.

2. A particle moves along a line according to the position equation $s(t) = t^4 - 4t^3$, where $t \geq 0$. Find all t for which the velocity is increasing.

3. A particle is moving along a horizontal line according to the equation $s(t) = 3t^2 - t^3$, where s meters is the directed distance of the particle from the origin at t seconds and $t \geq 0$. Find where the particle is speeding up.

4. A particle moves along a line according to the position equation $s(t) = \frac{1}{4}t^4 - 2t^3 + 4t^2$, where $0 \leq t \leq 4$. What is the total distance traveled by the particle?

(Answers are on page 153.)

Mean Value Theorem

The *Mean Value Theorem* states that if $f(x)$ is a continuous function on the closed interval $[a, b]$ and differentiable on the open interval (a, b), there is at least one point c in (a, b) such that $f'(c) = \frac{f(b) - f(a)}{b - a}$.

1+2=3 MATH TALK!

The Mean Value Theorem states that if a function is continuous and differentiable over an interval, there is at least one point in that interval where the instantaneous rate of change at that point is equal to the average rate of change over the interval. Graphically, it is stating that the slope of a tangent line is equal to the slope of a secant line at least once. This is shown in the figure below where there is at least one tangent line parallel to the secant line.

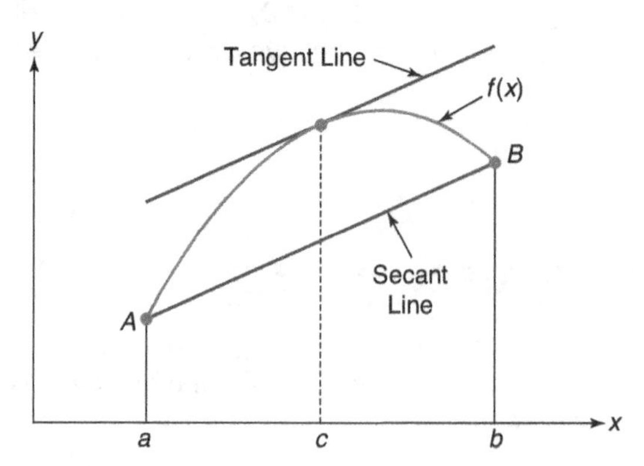

This theorem proves the existence of the first derivative based upon the function values. It can also prove the existence of the second derivative based upon the first derivative values.

Example 38:

Mike biked to work from home one morning. He did not make any stops along the 20-mile route and arrived 94 minutes later. What does the Mean Value Theorem say about his speed along the way?

Solution:

Since Mike made no stops along the way, his path was continuous. By the Mean Value Theorem, there was at least one point on the trip where the instantaneous rate of change was equal to the average rate of change. Therefore, at least once, Mike's speed was approximately $\frac{20-0}{94-0}=\frac{10}{47}$ miles/minute.

Example 39:

Let $f(x)=\sqrt{x}$. Find all values of c that satisfy the Mean Value Theorem on the closed interval $[0, 4]$.

Solution:

The given function is continuous over the given interval. Therefore, the Mean Value Theorem can be applied. In order to use the Mean Value Theorem, the derivative of the function and the average rate of change over the interval must be determined:

$$f'(x)=\frac{1}{2}x^{-1/2}=\frac{1}{2\sqrt{x}}$$

$$f'(c)=\frac{1}{2\sqrt{c}}$$

By the Mean Value Theorem:

$$f'(c)=\frac{f(b)-f(a)}{b-a}$$

$$f'(c)=\frac{f(4)-f(0)}{4-0}$$

$$\frac{1}{2\sqrt{c}}=\frac{\sqrt{4}-\sqrt{0}}{4-0}$$

$$4\sqrt{c}=4$$

$$\sqrt{c}=1$$

$$c=1$$

Example 40:

Selected values of $f'(x)$ are shown in the table below. The function $f(x)$ has a continuous second derivative for all real numbers. Does a value of x exist in the interval $[0, 6]$ such that $f''(x) = \frac{3}{2}$?

x	0	1	2	3	4	5	6
$f'(x)$	0.8	3.5	2	4	1.2	7	9.8

Solution:

Since the example is giving values about the first derivative and asking about the existence of the second derivative, the Mean Value Theorem should be used here. The question states that the function has a continuous second derivative. Therefore, the Mean Value Theorem can be applied:

$$f''(c) = \frac{f'(b) - f'(a)}{b - a}$$

$$f''(c) = \frac{f'(6) - f'(0)}{6 - 0}$$

$$f''(c) = \frac{9.8 - 0.8}{6 - 0}$$

$$f''(c) = \frac{9}{6}$$

$$f''(c) = \frac{3}{2}$$

By the Mean Value Theorem, there exists an x-value in the interval $[0, 6]$ such that $f''(x) = \frac{3}{2}$.

Related Rates

Related rates is an application of implicit differentiation. Related rates problems involve finding a rate at which a quantity changes by relating that quantity to other quantities whose rates of change are known. The rate of change is usually with respect to time.

Related rates problems are *painless* and involve seven steps.

Step 1: Define variables, starting with t. Include a diagram.

Step 2: Write down any numerical facts about the variables and their derivatives (rates) with respect to t. This is the information that you "know."

Step 3: Write down what you are being asked to find. This is what you "want."

Step 4: Write an equation that relates all of the variables. Refer to your diagram.

Step 5: Differentiate both sides of the equation with respect to t.

Step 6: Substitute your "known" information, and solve for what you "want."

Step 7: Write a concluding sentence about your solution, including direction and any other information that was asked for.

Example 41:

Keira leans a 25-foot ladder against a wall. Keira steps away, and the bottom of the ladder starts to slip away from the wall at a rate of 3 feet/second. How fast is the top of the ladder sliding down when the bottom of the ladder is 15 feet from the wall?

Solution:

Step 1: Define variables, starting with t. Include a diagram.

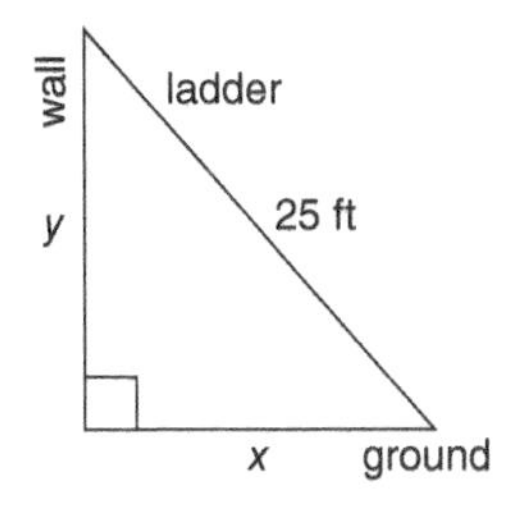

Let t = time in seconds

Let x = distance between wall and bottom of the ladder in feet

Let y = distance between ground and top of the ladder in feet

Step 2: Write down any numerical facts about the variables and their derivatives (rates) with respect to t. This is the information that you "know."

Know: $\frac{dx}{dt} = +3$ ft/sec

This is from the second sentence; it is positive because the foot of the ladder moves away from the wall and the distance between the two gets larger.

Step 3: Write down what you are being asked to find. This is what you "want."

Want: $\frac{dy}{dt}$ when $x = 15$ ft

This is from the last sentence.

Step 4: Write an equation that relates all of the variables. Refer to your diagram.

It is assumed that the wall is perpendicular to the ground, forming a right angle. Since the diagram is a right triangle, an equation that relates all of the variables is the Pythagorean Theorem:

$$x^2 + y^2 = 25^2$$

Step 5: Differentiate both sides of the equation with respect to t.

Use implicit differentiation to differentiate the equation with respect to t:

$$\frac{d}{dt}(x^2 + y^2) = \frac{d}{dt}(25^2)$$

$$2x\frac{dx}{dt} + 2y\frac{dy}{dt} = 0$$

$$x\frac{dx}{dt} + y\frac{dy}{dt} = 0$$

Step 6: Substitute your "known" information, and solve for what you "want."

It is known that $\frac{dx}{dt} = +3$ ft/sec, and you want $\frac{dy}{dt}$ when $x = 15$ ft. The only other variable that is still unknown is y. To solve for y at this particular instant, first substitute the known information into the original equation:

$$x^2 + y^2 = 25^2$$

$$(15)^2 + y^2 = 25^2$$

$$y = 20$$

Then substitute all known information into the derivative:

$$x\frac{dx}{dt} + y\frac{dy}{dt} = 0$$

$$(15)(3) + (20)\frac{dy}{dt} = 0$$

$$\frac{dy}{dt} = -2.25 \text{ ft/sec}$$

Step 7: Write a concluding sentence about your solution, including direction and any other information that was asked for.

When the bottom of the ladder is 15 feet from the wall, the top of the ladder is sliding down at a rate of 2.25 feet/second.

1+2=3 MATH TALK!

When the rate of a quantity (or variable) is a positive value, the quantity (or variable) is increasing. When the rate of a quantity (or variable) is a negative value, the quantity (or variable) is decreasing. When writing your concluding sentence, be careful choosing what words you use to describe the rates. If you are including direction, do not include the sign of the value since the direction, such as down, will already indicate that the rate is negative.

CAUTION—Major Mistake Territory!

Be careful to not include all the information right away when setting up an equation to model the related rates problem. Constants should be included in an equation in Step 2 only if it is a known constant, a value that is never changing. In the previous example, the 25 feet was included right away because the ladder is a fixed length and unchanging. If the distance is known to change (get bigger or smaller), it must be represented by a variable. Notice that the $x = 15$ and $y = 20$ were substituted only at the end of the problem since these were not constant lengths and represented a distance at only a particular time. Substituting in constants too soon will result in derivatives of zero and will eliminate important parts of the equation.

A common related rates problem in calculus involves cones. Often in problems such as the one below, proportions will have to be used to write the equation in terms of fewer variables.

Example 42:

Water is flowing at a rate of 2 meter³/minute into a tank in the form of an inverted cone that has an altitude of 16 meters and a radius of 4 meters. How fast is the water level rising when the water is 5 meters deep?

Solution:

Step 1: Define variables, starting with t. Include a diagram.

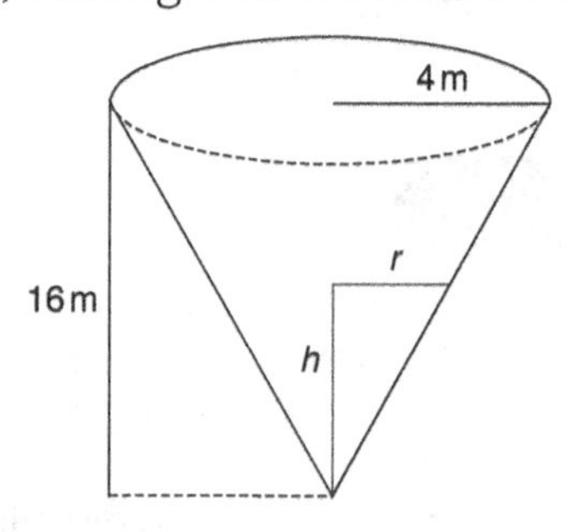

Let $t =$ time in minutes

Let $h =$ height of water in meters

Let $r =$ radius of water in meters

Let $V =$ volume of water in cubic meters

Step 2: Write down any numerical facts about the variables and their derivatives (rates) with respect to t. This is the information that you "know."

Know: $\frac{dV}{dt} = +2 \text{ m}^3/\text{min}$

This is from the first sentence; it is positive because the cone is being filled with water.

Step 3: Write down what you are being asked to find. This is what you "want."

Want: $\frac{dh}{dt}$ when $h = 5$ m

This is from the last sentence.

Step 4: Write an equation that relates all of the variables. Refer to your diagram.

Since the diagram is a cone, an equation that relates all of the variables is the formula for the volume of a cone:

$$V = \frac{1}{3}\pi r^2 h$$

Step 5: Differentiate both sides of the equation with respect to t.

Use implicit differentiation and the Product Rule to differentiate the equation with respect to t:

$$\frac{d}{dt}(V) = \frac{d}{dt}\left(\frac{1}{3}\pi r^2 h\right)$$

$$\frac{dV}{dt} = \frac{1}{3}\pi r^2 \bullet \frac{dh}{dt} + h \bullet \frac{1}{3}\pi \bullet 2r\frac{dr}{dt}$$

Step 6: Substitute your "known" information, and solve for what you "want."

It is known that $\frac{dV}{dt} = +2 \text{ m}^3/\text{min}$, and you want $\frac{dh}{dt}$ when $h = 5$ m. The other unknown variables are r and $\frac{dr}{dt}$. Unlike the previous example with only one other unknown variable, this example has too many unknown variables. Step 4 needs to be revisited and a different equation needs to be written that is in terms of one variable.

Step 4 revisited: Write an equation that relates all of the variables. Refer to your diagram. In the diagram are two right triangles, one formed by the dimensions of the cone and one formed by the dimensions of the water.

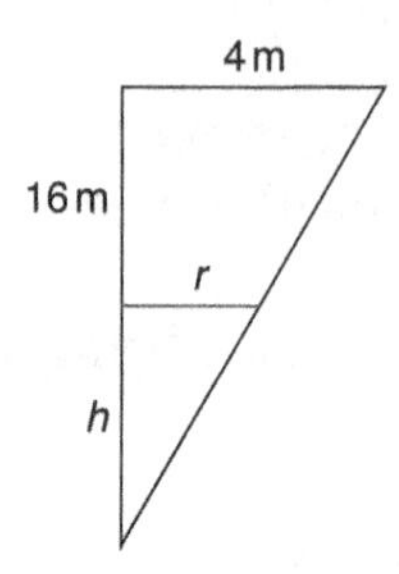

These are similar triangles and, as such, their corresponding sides are in proportion:

$$\frac{16}{4} = \frac{h}{r}$$

Since the problem wants $\frac{dh}{dt}$, solve for r in terms of h:

$$r = \frac{h}{4}$$

Substitute into the volume equation and simplify:

$$V = \frac{1}{3}\pi\left(\frac{h}{4}\right)^2 h = \frac{1}{3}\pi\frac{h^2}{16}h$$

$$V = \frac{\pi}{48}h^3$$

Step 5 revisited: Differentiate both sides of the equation with respect to t.

Use implicit differentiation to differentiate the equation with respect to t:

$$\frac{d}{dt}(V) = \frac{d}{dt}\left(\frac{\pi}{48}h^3\right)$$

$$\frac{dV}{dt} = \frac{\pi}{48} \bullet 3h^2 \bullet \frac{dh}{dt}$$

$$\frac{dV}{dt} = \frac{\pi}{16}h^2 \bullet \frac{dh}{dt}$$

Step 6 revisited: Substitute your "known" information, and solve for what you "want." It is known that $\frac{dV}{dt} = +2\ \text{m}^3/\text{min}$, and you want $\frac{dh}{dt}$ when $h = 5$ m. There are no other unknown variables. Substitute all known information into the derivative:

$$\frac{dV}{dt} = \frac{\pi}{16}h^2 \bullet \frac{dh}{dt}$$

$$(2) = \frac{\pi}{16}(5)^2 \bullet \frac{dh}{dt}$$

$$\frac{dh}{dt} = \frac{32}{25\pi}\ \text{m/min}$$

Step 7: Write a concluding sentence about your solution, including direction and any other information that was asked for.

When the water is 5 meters deep, the water level is rising at a rate of $\frac{32}{25\pi}$ meters/minute.

Example 43:

Two cars are traveling toward the intersection of two roads. Car 1 is going due east at a rate of 90 kilometers/hour, and car 2 is going due south at a rate of 60 kilometers/hour. At what rate are the cars approaching each other at the instant when car 1 is 0.2 kilometers and car 2 is 0.15 kilometers from the intersection?

Solution:

Step 1: Define variables, starting with t. Include a diagram.

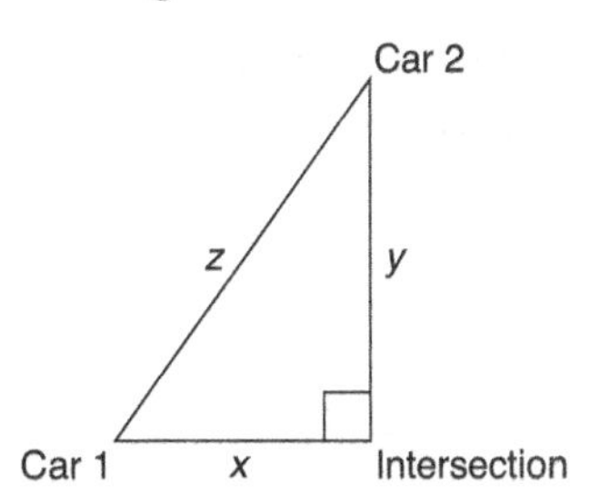

Let t = time in hours

Let x = distance car 1 is from the intersection in kilometers

Let y = distance car 2 is from the intersection in kilometers

Let z = distance between car 1 and car 2 in kilometers

Step 2: Write down any numerical facts about the variables and their derivatives (rates) with respect to t. This is the information that you "know."

Know: $\frac{dx}{dt} = -90$ km/hr and $\frac{dy}{dt} = -60$ km/hr

This is from the first sentence; both rates are negative since the distance between each car as they approach the intersection is getting smaller.

Step 3: Write down what you are being asked to find. This is what you "want."

Want: $\frac{dz}{dt}$ when $x = 0.2$ km and $y = 0.15$ km

This is from the last sentence.

Step 4: Write an equation that relates all of the variables. Refer to your diagram.

Since the diagram is a right triangle, an equation that relates all of the variables is the Pythagorean Theorem:

$$x^2 + y^2 = z^2$$

Step 5: Differentiate both sides of the equation with respect to t.

Use implicit differentiation to differentiate the equation with respect to t:

$$\frac{d}{dt}(x^2 + y^2) = \frac{d}{dt}(z^2)$$

$$2x\frac{dx}{dt} + 2y\frac{dy}{dt} = 2z\frac{dz}{dt}$$

$$x\frac{dx}{dt} + y\frac{dy}{dt} = z\frac{dz}{dt}$$

Step 6: Substitute your "known" information, and solve for what you "want."

It is known that $\frac{dx}{dt} = -90$ km/hr and $\frac{dy}{dt} = -60$ km/hr. You want $\frac{dz}{dt}$ when $x = 0.2$ km and $y = 0.15$ km. The only other variable that is still unknown is z. To solve for z at this particular instant, first substitute the information into the original equation:

$$x^2 + y^2 = z^2$$

$$(0.2)^2 + (0.15)^2 = z^2$$

$$z = 0.25$$

Then substitute all known information into the derivative:

$$x\frac{dx}{dt} + y\frac{dy}{dt} = z\frac{dz}{dt}$$

$$(0.2)(-90) + (0.15)(-60) = (0.25)\frac{dz}{dt}$$

$$\frac{dz}{dt} = -108 \text{ km/hr}$$

Step 7: Write a concluding sentence about your solution, including direction and any other information that was asked for.

When car 1 is 0.2 kilometers from the intersection and car 2 is 0.15 kilometers from the intersection, the rate at which the cars approach each other is –108 kilometers/hour.

1+2=3 MATH TALK!

Notice in Step 7 of Example 43 that the negative sign was included in the rate. That is because, in this example, no descriptive word was used to indicate the rate at which the two cars approached each other.

BRAIN TICKLERS Set # 22

1. Let f be the function given by $f(x) = x^3 - 4x$. Find all values of c that satisfy the conclusion of the Mean Value Theorem on the closed interval [0, 4].

2. The radius r of a sphere is increasing at a constant rate of 0.3 inches per second. Find the rate of increase, in cubic inches per second, in the volume V at the instant when the surface area S becomes 100π inches^2. (Hint: $S = 4\pi r^2$ and $V = \frac{4}{3}\pi r^3$.)

3. A train track and a road cross at right angles. Lawrence stands on the road 60 meters south of the intersection and watches an eastbound train traveling at 50 meters/second. At how many meters per second, to the nearest tenth, is the train moving away from Lawrence 2 seconds after it passes through the intersection?

4. A right inverted cone with a height of 10 centimeters and a radius of 5 centimeters is filled with water. The water in the cone is leaking out of the bottom so that the depth of the water, h, is changing at a constant rate of $-\frac{3}{10}$ centimeters/hour. Find the rate of change of the volume of water in the container, with respect to time, when $h = 5$ centimeters.

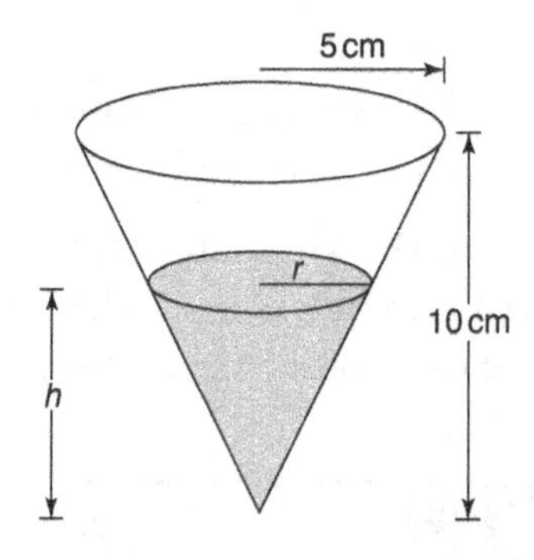

(Answers are on page 154.)

BRAIN TICKLERS—THE ANSWERS

Set # 17, page 92

1. $y - 1 = -4(x + 1)$
2. $y - 3 = \frac{1}{e^3}(x - e^3)$
3. $y + 1 = \frac{1}{6}(x - 1)$
4. $\frac{1}{20}$

Set # 18, page 99

1. 3
2. 0
3. 1
4. 1

Set # 19, page 118

1. $(0, 2)$
2. $x = -1$
3. $(-\sqrt{2}, 0) \cup (\sqrt{2}, \infty)$
4. $(3, 162)$

Set # 20, page 131

1. Relative maximum at $x = -2$; relative minimum at $x = 0$; point of inflection at $x = -1, 1, 2$
2. $f'(c) = 0$ and $f''(c) < 0$
3. $\frac{130}{3}$
4. $\frac{10}{3}$ inch by $\frac{10}{3}$ inch

Set # 21, pages 139–140

1. The particle is moving to the right when t is in the intervals $[0, 1) \cup (2, \infty)$.
2. The velocity is increasing $(a(t) > 0)$ when t is in the interval $(2, \infty)$.
3. The particle is speeding up when t is in the intervals $[0, 1) \cup (2, \infty)$.
4. The total distance traveled by the particle is 8 units.

Set # 22, page 152

1. $c = +\frac{2}{\sqrt{3}}$
2. $\frac{dV}{dt} = 30\pi$ in.^{3}/sec
3. 42.9 m/sec
4. $\frac{dV}{dt} = -\frac{15\pi}{8}$ cm^3/hr

Chapter 5

Antidifferentiation

Like the title of the chapter suggests, a simple way of defining *antidifferentiation* is that it is the opposite of taking a derivative. The antiderivative is the operation that goes backward from the derivative of a function to the function itself.

To understand how to find the antiderivative, first consider the following functions and their derivatives.

Functions $f(x)$ and $g(x)$	Derivative Functions $f'(x)$ and $g'(x)$
$f(x) = x^3 + 7$	$f'(x) = 3x^2$
$f(x) = x^3 - 2$	$f'(x) = 3x^2$
$f(x) = x^3$	$f'(x) = 3x^2$
$g(x) = \sin(x) + \frac{\pi}{2}$	$g'(x) = \cos(x)$
$g(x) = \sin(x) - 10$	$g'(x) = \cos(x)$
$g(x) = \sin(x)$	$g'(x) = \cos(x)$

The derivatives for $f(x)$ and the derivatives for $g(x)$ are all the same even though the original functions were different from each other. The original functions differed in their constant terms. This needs to be taken into consideration when performing antidifferentiation. The more common name for the antiderivative is the *indefinite integral.*

Indefinite Integrals

The notation for an indefinite integral is shown in the table below.

Notation	Read as
$\int f(x)\,dx = F(x) + C$	The indefinite integral of $f(x)$ with respect to x equals $F(x)$ plus C
$\int$	Integral symbol
$f(x)$	Integrand
C	Constant

The term *indefinite* is used because the right side of the equation is not a definite function but, instead, a whole possibility of functions. That is why all solutions to any indefinite integral must include a constant or "plus C."

PAINLESS TIP

When evaluating an integral, it is important that it is with respect to the same variable that is shown in the integrand. For example, if the integrand is in terms of x, you must include dx. If the integrand is in terms of y, you must include dy. It is extremely important to always include the dx or dy in your integral notation; otherwise, evaluating an integral is impossible.

Since the indefinite integral is how to find the antiderivatives of expressions, most indefinite integrals can be determined by understanding derivatives. For example, since the derivative of $\sin(x)$ is $\cos(x)$, then working backward, the antiderivative of $\cos(x)$ will involve $\sin(x)$. More formally, $\int \cos(x)\,dx = \sin(x) + C$. When finding the indefinite integral, you can also take the derivative of your solution to check that it matches the integrand.

For polynomial functions, the formula below can be followed for a given monomial, where a and n are real numbers:

$$\int ax^n\,dx = a \bullet \frac{x^{n+1}}{n+1} + C$$

1+2=3 MATH TALK!

This formula is the reverse Power Rule for derivatives. When integrating, add 1 to the exponent and divide by the new exponent. This rule can be applied to each term in a polynomial expression.

Example 1:

Evaluate $\int 5x^{12}dx$.

Solution:

$$\int 5x^{12}\,dx = 5 \bullet \frac{x^{12+1}}{12+1} = 5 \bullet \frac{x^{13}}{13} + C$$

To check, take the derivative of the answer to see if it matches the integrand:

$$\frac{d}{dx}\left(5 \bullet \frac{x^{13}}{13} + C\right) = 5 \bullet \frac{13x^{12}}{13} + 0 = 5x^{12}$$

Example 2:

Evaluate $\int \sqrt{x}\,dx$.

Solution:

Rewrite the integrand using exponents:

$$\int \sqrt{x}\,dx = \int x^{1/2}\,dx$$

$$\int x^{1/2}\,dx = \frac{x^{1/2+1}}{\frac{1}{2}+1} + C = \frac{x^{3/2}}{\frac{3}{2}} + C = \frac{2}{3}\sqrt{x^3} + C$$

Example 3:

Evaluate $\int x^{-2}dx$.

Solution:

$$\int x^{-2}\,dx = \frac{x^{-2+1}}{-2+1} + C = -x^{-1} + C = -\frac{1}{x} + C$$

Properties of Indefinite Integrals

Three integral properties can be applied to the integral shortcut rules.

Property 1: The integral of a constant is equal to the constant multiplied by the variable plus C.

If a is a constant, $\int a\,dx = ax + C$.

Property 2: A constant factor can be moved through an integral sign.

If a is a constant, then $\int a f(x)\,dx = a\int f(x)\,dx$.

Property 3: An antiderivative of a sum or difference of two or more differentiable functions is equal to the sum or difference of their antiderivatives.

$$\int (f(x) \pm g(x))dx = \int f(x)dx \pm \int g(x)dx$$

Example 4:

Evaluate the following integrals:

1. $\int (x^{2/3} - 4x^{-1/5} + 4)\,dx$
2. $\int (2 + y^2)^2 dy$
3. $\int \frac{1 - 2t^3}{t^3}\,dt$

Solution:

1. Use Property 3:

$$\int (x^{2/3} - 4x^{-1/5} + 4)dx = \int x^{2/3}dx - \int 4x^{-1/5}dx + \int 4dx$$

Use Property 2:

$$\int x^{2/3}dx - \int 4x^{-1/5}dx + \int 4dx = \int x^{2/3}dx - 4\int x^{-1/5}dx + \int 4dx$$

By Property 1 and the rule for monomials:

$$\int x^{2/3}dx - 4\int x^{-1/5}dx + \int 4dx = \frac{x^{2/3+1}}{\frac{2}{3}+1} - 4 \bullet \frac{x^{-1/5+1}}{-\frac{1}{5}+1} + 4x + C$$

$$= \frac{x^{5/3}}{\frac{5}{3}} - 4 \bullet \frac{x^{4/5}}{\frac{4}{5}} + 4x + C$$

$$= \frac{3x^{5/3}}{5} - \frac{4 \bullet 5x^{4/5}}{4} + 4x + C$$

$$= \frac{3x^{5/3}}{5} - 5x^{4/5} + 4x + C$$

2. Expand the binomial:

$$\int (2 + y^2)^2 dy = \int (2 + y^2)(2 + y^2)dy = \int (4 + 4y^2 + y^4)dy$$

Use Properties 2 and 3:

$$\int (4 + 4y^2 + y^4)dy = \int 4dy + 4\int y^2 dy + \int y^4 dy$$

By Property 1 and the rule for monomials:

$$\int 4dy + 4\int y^2 dy + \int y^4 dy = 4y + 4 \bullet \frac{y^{2+1}}{2+1} + \frac{y^{4+1}}{4+1} + C$$

$$= 4y + \frac{4y^3}{3} + \frac{y^5}{5} + C$$

3. Divide each term in the numerator by the denominator and use Property 3:

$$\int \frac{1 - 2t^3}{t^3} dt = \int \frac{1}{t^3} dt - \int \frac{2t^3}{t^3} dt$$

By simplifying and rewriting the exponents, the integral can be evaluated using Property 1 and the rule for monomials:

$$\int \frac{1}{t^3}dt - \int \frac{2t^3}{t^3}dt = \int t^{-3}dt - \int 2dt$$
$$= \frac{t^{-3+1}}{-3+1} - 2t + C$$
$$= \frac{t^{-2}}{-2} - 2t + C$$
$$= -\frac{1}{2t^2} - 2t + C$$

Integral Formulas

The tables below list important indefinite integrals to remember.

Trigonometric Integration Formulas		
$\int \cos(x)dx = \sin(x) + C$	$\int \sin(x)dx = -\cos(x) + C$	$\int \sec^2(x)dx = \tan(x) + C$
$\int \csc^2(x)dx = -\cot(x) + C$	$\int \sec(x)\tan(x)dx = \sec(x) + C$	$\int \csc(x)\cot(x)dx = -\csc(x) + C$

Exponential and Logarithmic Integration Formulas		
$\int e^x dx = e^x + C$	$\int b^x dx = \frac{b^x}{\ln(b)} + C$	$\int \frac{1}{x}dx = \ln\lvert x\rvert + C$

Inverse Trigonometric Integration Formulas		
$\int \frac{1}{1+x^2}dx = \tan^{-1}(x) + C$	$\int \frac{1}{\sqrt{1-x^2}}dx = \sin^{-1}(x) + C$	$\int \frac{1}{x\sqrt{x^2-1}}dx = \sec^{-1}\lvert x\rvert + C$
$\int -\frac{1}{1+x^2}dx = \cot^{-1}(x) + C$	$\int -\frac{1}{\sqrt{1-x^2}}dx = \cos^{-1}(x) + C$	$\int -\frac{1}{x\sqrt{x^2-1}}dx = \csc^{-1}\lvert x\rvert + C$

1+2=3 **MATH TALK!**

The exponential function, e^x, is interesting because the derivative and indefinite integral are the same function, e^x. It is the only function in calculus to behave that way.

Example 5:

Evaluate $\int (4\sec^2(x) + \csc(x)\cot(x))\,dx$.

Solution:

By Properties 2 and 3:

$$\int (4\sec^2(x) + \csc(x)\cot(x))\,dx = 4\int \sec^2(x)\,dx + \int \csc(x)\cot(x)\,dx$$
$$= 4\tan(x) - \csc(x) + C$$

Example 6:

Evaluate $\int 3^x\,dx$.

Solution:

Use the exponential integration formula:

$$\int 3^x\,dx = \frac{3^x}{\ln(3)} + C$$

Example 7:

Evaluate $\int \frac{dx}{2x}$.

Solution:

By Property 1:

$$\int \frac{dx}{2x} = \frac{1}{2}\int \frac{1}{x}\,dx$$

Use the logarithmic integration formula:

$$\frac{1}{2}\int \frac{1}{x}\,dx = \frac{1}{2}\ln|x| + C$$

CAUTION—Major Mistake Territory!

Be careful when you encounter $\int \frac{1}{x}dx$. It is easy to forget the formula and try evaluating the integral by rewriting using exponents and using the monomial rule:

$$\int x^{-1}dx = \frac{x^{-1+1}}{-1+1} + C = \frac{1}{0} + C$$

Following this logic, the integral is undefined.
Therefore, $\int \frac{1}{x}dx \neq \int x^{-1}\,dx$.
Avoid this by recognizing that $\int \frac{1}{x}dx = \ln|x| + C$ since $\frac{d}{dx}(\ln(x)) = \frac{1}{x}$.

BRAIN TICKLERS Set # 23

1. Evaluate $\int \left(t^2 - \frac{1}{t^2}\right)dt$.
2. Evaluate $\int \csc^2(x)\,dx$.
3. Evaluate $\int \frac{1}{2x}dx$.
4. Evaluate $\int \frac{3}{1+x^2}dx$.

(Answers are on page 207.)

Like differentiation, additional rules are sometimes needed to evaluate an integral. The following are different tools that can be applied to antidifferentiation with an explanation on when to use them.

Integration by *U*-Substitution

U-substitution is an integration tool that allows you to rewrite an integral into a more recognizable form. Use this technique when an integral is very similar to one of the integration formulas or if you must expand an expression raised to a power that is greater than 2.

U-substitution is *painless* and involves four steps.

Step 1: Let u represent part of the integrand. This is usually the input of a function or an expression that, when differentiated, appears elsewhere in the integrand.

Step 2: Differentiate u with respect to x or with respect to the variable it is written in. Then solve for dx.

Step 3: Rewrite the integral in terms of u and evaluate.

Step 4: After evaluating, substitute back in for u.

Example 8:

Evaluate $\int (x+3)^{15}\,dx$.

Solution:

Since the integrand is a binomial raised to a power greater than 2, try u-substitution.

Step 1: Let u represent part of the integrand.

Let $u = x + 3$, the input of the power function.

Step 2: Differentiate u with respect to x:

$$\frac{d}{dx}(u) = \frac{d}{dx}(x+3)$$

$$\frac{du}{dx} = 1$$

Solve for dx:

$$du = 1 \bullet dx \rightarrow dx = du$$

Step 3: Rewrite the integral in terms of u and evaluate:

$$\int (x+3)^{15} dx \rightarrow \int u^{15} du = \frac{u^{15+1}}{15+1} + C = \frac{u^{16}}{16} + C$$

Step 4: After evaluating, substitute back in for u:

$$\frac{u^{16}}{16} + C = \frac{(x+3)^{16}}{16} + C$$

Example 9:

Evaluate $\int (4x+3)^{15}dx$.

Solution:

Since the integrand is a binomial raised to a power greater than 2, try u-substitution.

Step 1: Let u represent part of the integrand.

Let $u = 4x + 3$, the input of the power function.

Step 2: Differentiate u with respect to x:

$$\frac{d}{dx}(u) = \frac{d}{dx}(4x+3)$$

$$\frac{du}{dx} = 4$$

Solve for dx:

$$du = 4 \bullet dx \rightarrow dx = \frac{1}{4}du$$

Step 3: Rewrite the integral in terms of u and evaluate:

$$\int (4x+3)^{15}dx \rightarrow \int u^{15} \bullet \frac{1}{4}du = \frac{1}{4}\int u^{15}du$$

$$= \frac{1}{4} \bullet \frac{u^{15+1}}{15+1} + C = \frac{u^{16}}{64} + C$$

Step 4: After evaluating, substitute back in for u:

$$\frac{u^{16}}{64} + C = \frac{(4x+3)^{16}}{64} + C$$

1+2=3 MATH TALK!

Example 9 differs from Example 8 in Step 2 since Example 9 has the additional constant. Often it is easy enough to rewrite the new integral by substituting in for the additional constant. Other times, you will choose u based upon what its derivative will be. For example, if the integrand is a product or quotient, there are two terms. Typically, the u-term will be the input of the more complicated term and its derivative will be like the other term in the integrand. This will make it easier to rewrite the integral completely in terms of u.

Example 10:
Evaluate $\int x^2(x^3 + 3)^{15}dx$.

Solution:

In this example, two terms are being multiplied in the integrand. This is another way to recognize u-substitution. If the derivative of the u-term matches the other term in the integrand, u-substitution is the technique that should be used.

Step 1: Let u represent part of the integrand.

Let $u = x^3 + 3$, the input of the power function.

Step 2: Differentiate u with respect to x. Then solve for dx:

$$\frac{d}{dx}(u) = \frac{d}{dx}(x^3 + 3)$$

$$\frac{du}{dx} = 3x^2$$

The derivative has an x^2, which is the other term in the integrand. We can make a substitution for this. Solve for x^2dx:

$$du = 3x^2dx \rightarrow x^2dx = \frac{1}{3}du$$

Step 3: Rewrite the integral in terms of u and evaluate:

$$\int x^2(x^3 + 3)^{15}dx = \int (x^3 + 3)^{15}x^2dx \rightarrow \int u^{15} \bullet \frac{1}{3}du$$

$$\int u^{15} \bullet \frac{1}{3}du = \frac{1}{3}\int u^{15}du = \frac{1}{3} \bullet \frac{u^{15+1}}{15+1} + C = \frac{u^{16}}{48} + C$$

Step 4: After evaluating, substitute back in for u:

$$\frac{u^{16}}{48} + C = \frac{(x^3 + 3)^{16}}{48} + C$$

Example 11:
Evaluate $\int \frac{x^2}{x^3 + 4}dx$.

Solution:

In this example, two terms are being divided in the integrand. This is another way to recognize u-substitution. If the derivative of the u-term matches the other term in the integrand, u-substitution is the technique that should be used.

Step 1: Let u represent part of the integrand.

Let $u = x^3 + 4$ since this is in the denominator and substituting for the derivative in the numerator is easier.

Step 2: Differentiate u with respect to x:

$$\frac{d}{dx}(u) = \frac{d}{dx}(x^3 + 4)$$

$$\frac{du}{dx} = 3x^2$$

The derivative has an x^2, which is the other term in the integrand. We can make a substitution for this instead. Solve for x^2dx:

$$du = 3x^2dx \rightarrow x^2dx = \frac{1}{3}du$$

Step 3: Rewrite the integral in terms of u and evaluate:

$$\int \frac{x^2}{x^3 + 4}dx = \int \frac{1}{x^3 + 4}x^2dx \rightarrow \int \frac{1}{u} \bullet \frac{1}{3}du$$

$$\frac{1}{3}\int \frac{1}{u}du = \frac{1}{3}\ln|u| + C$$

Step 4: After evaluating, substitute back in for u:

$$\frac{1}{3}\ln|u| + C = \frac{1}{3}\ln|x^3 + 4| + C$$

Example 12:

Evaluate $\int e^{5x}dx$.

Solution:

Since this integral is very similar to one of the integration formulas, $\int e^x dx$, try u-substitution.

Step 1: Let u represent part of the integrand.

Let $u = 5x$, the input of the exponential function.

Step 2: Differentiate u with respect to x:

$$\frac{d}{dx}(u) = \frac{d}{dx}(5x)$$

$$\frac{du}{dx} = 5$$

Solve for dx:

$$du = 5 \bullet dx \rightarrow dx = \frac{1}{5}du$$

Step 3: Rewrite the integral in terms of u and evaluate:

$$\int e^{5x}dx \rightarrow \int e^{u} \bullet \frac{1}{5}du = \frac{1}{5}\int e^{u}du = \frac{1}{5} \bullet e^{u} + C$$

Step 4: After evaluating, substitute back in for u:

$$\frac{1}{5} \bullet e^{u} + C = \frac{1}{5}e^{5x} + C$$

Example 13:

Evaluate $\int \cos(x)2^{\sin(x)}dx$.

Solution:

In this example, two terms are being multiplied in the integrand. This is another way to recognize u-substitution. If the derivative of the u-term matches the other term in the integrand, u-substitution is the technique that should be used.

Step 1: Let u represent part of the integrand.

Let $u = \sin(x)$, the input of the exponential function.

Step 2: Differentiate u with respect to x:

$$\frac{d}{dx}(u) = \frac{d}{dx}(\sin(x))$$

$$\frac{du}{dx} = \cos(x)$$

The derivative is $\cos(x)$, which is the other term in the integrand. We can make a substitution for this instead. Solve for $\cos(x)dx$:

$$du = \cos(x)dx \rightarrow \cos(x)dx = du$$

Step 3: Rewrite the integral in terms of u and evaluate:

$$\int \cos(x)2^{\sin(x)}dx = \int 2^{\sin(x)} \bullet \cos(x)dx \rightarrow \int 2^u du$$

$$\int 2^u du = \frac{2^u}{\ln(2)} + C$$

Step 4: After evaluating, substitute back in for u:

$$\frac{2^u}{\ln(2)} + C = \frac{2^{\sin(x)}}{\ln(2)} + C$$

Example 14:

Evaluate $\int \tan(x)\, dx$.

Solution:

This does not match any of the integral formulas. This expression can be rewritten as a quotient of two trigonometric expressions. This is another way to recognize u-substitution. If the derivative of the u-term matches the other term in the integrand, u-substitution is the technique that should be used:

$$\int \tan(x)dx = \int \frac{\sin(x)}{\cos(x)}dx$$

Step 1: Let u represent part of the integrand.

Let $u = \cos(x)$ since this is in the denominator and substituting for the derivative in the numerator is easier.

Step 2: Differentiate u with respect to x:

$$\frac{d}{dx}(u) = \frac{d}{dx}(\cos(x))$$

$$\frac{du}{dx} = -\sin(x)$$

The derivative has $\sin(x)$, which is the other term in the integrand. We can make a substitution for this instead. Solving for $\sin(x)dx$:

$$du = -\sin(x)dx \rightarrow \sin(x)dx = -1du$$

Step 3: Rewrite the integral in terms of u and evaluate:

$$\int \tan(x)\,dx = \int \frac{\sin(x)}{\cos(x)}\,dx = \int \frac{1}{\cos(x)} \bullet \sin(x)\,dx \rightarrow \int \frac{1}{u} \bullet -1\,du$$

$$\int \frac{1}{u} \bullet -1du = -1\int \frac{1}{u}du = -\ln|u| + C$$

Step 4: After evaluating, substitute back in for u:

$$-\ln|u| + C = -\ln|\cos(x)| + C$$

Example 15:

Evaluate $\int \sec(x)\,dx$.

Solution:

This does not match any of the integral formulas. Since this expression can be rewritten only as a quotient of one trigonometric expression, we will multiply it by a form of 1 to create an equivalent expression:

$$\int \sec(x)\left(\frac{\sec(x) + \tan(x)}{\sec(x) + \tan(x)}\right)dx$$

$$\int \frac{\sec^2(x) + \sec(x)\tan(x)}{\sec(x) + \tan(x)}dx$$

When written in this form, u-substitution can be recognized since the derivative of the denominator is the numerator.

Step 1: Let u represent part of the integrand.

Let $u = \sec(x) + \tan(x)$ since this is in the denominator and substituting for the derivative in the numerator is easier.

Step 2: Differentiate u with respect to the x, or with respect to the variable it is written in. Then solve for dx.

Take the derivative of u with respect to x:

$$\frac{d}{dx}(u) = \frac{d}{dx}(\sec(x) + \tan(x))$$

$$\frac{du}{dx} = \sec(x)\tan(x) + \sec^2(x) \rightarrow du = (\sec(x)\tan(x) + \sec^2(x))dx$$

This expression appears in the numerator of the integrand.

Step 3: Rewrite the integral in terms of u and evaluate:

$$\int \sec(x)dx = \int \frac{\sec^2(x) + \sec(x)\tan(x)}{\sec(x) + \tan(x)}dx \rightarrow \int \frac{du}{u} = \ln|u| + C$$

Step 4: After evaluating, substitute back in for u:

$$\ln|u| + C = \ln|\sec(x) + \tan(x)| + C$$

BRAIN TICKLERS Set # 24

1. Evaluate $\int \frac{4x^2}{(1-8x^3)^4}dx$.
2. Evaluate $\int \sin x \sin(\cos x)\, dx$.
3. Evaluate $\int \frac{\sec^2 x}{\tan x}dx$.
4. Evaluate $\int \frac{5x^4}{x^5+1}dx$.

(Answers are on page 207.)

Sometimes the integrand needs to be manipulated before u-substitution can be performed. This is usually when working with inverse trigonometric integration formulas.

Example 16:

Evaluate $\int \frac{1}{1+4x^2}\,dx$.

Solution:

Since this integral is very similar to one of the inverse trigonometric integration formulas, $\int \frac{1}{1+x^2}\,dx$, first manipulate the squared term and then try u-substitution:

$$\int \frac{1}{1+4x^2}\,dx = \int \frac{1}{1+(2x)^2}\,dx$$

Step 1: Let u represent part of the integrand.

Let $u = 2x$, the input of the squaring function.

Step 2: Differentiate u with respect to x:

$$\frac{d}{dx}(u) = \frac{d}{dx}(2x)$$

$$\frac{du}{dx} = 2$$

Solve for dx:

$$du = 2 \bullet dx \rightarrow dx = \frac{1}{2}du$$

Step 3: Rewrite the integral in terms of u and evaluate:

$$\int \frac{1}{1+(2x)^2}\,dx \rightarrow \int \frac{1}{1+u^2} \bullet \frac{1}{2}du = \frac{1}{2}\int \frac{1}{1+u^2}\,du$$

$$= \frac{1}{2}\tan^{-1}(u) + C$$

Step 4: After evaluating, substitute back in for u:

$$\frac{1}{2}\tan^{-1}(u) + C = \frac{1}{2}\tan^{-1}(2x) + C$$

Example 17:

Evaluate $\int \frac{1}{9+x^2}\,dx$.

Solution:

Since this integral is very similar to one of the inverse trigonometric integration formulas, $\int \frac{1}{1+x^2}\,dx$, first factor out a common perfect square, manipulate the squared term, and then try u-substitution:

$$\int \frac{1}{9+x^2}\,dx = \int \frac{1}{9\left(1+\frac{x^2}{9}\right)}\,dx = \frac{1}{9}\int \frac{1}{1+\left(\frac{x}{3}\right)^2}\,dx$$

Step 1: Let u represent part of the integrand.
Let $u = \frac{x}{3}$, the input of the squaring function.

Step 2: Differentiate u with respect to x:

$$\frac{d}{dx}(u) = \frac{d}{dx}\left(\frac{x}{3}\right)$$

$$\frac{du}{dx} = \frac{1}{3}$$

Solving for dx:

$$du = \frac{1}{3} \bullet dx \rightarrow dx = 3du$$

Step 3: Rewrite the integral in terms of u and evaluate:

$$\frac{1}{9}\int \frac{1}{1+\left(\frac{x}{3}\right)^2}\,dx \rightarrow \frac{1}{9}\int \frac{1}{1+u^2} \bullet 3du = \frac{1}{3}\int \frac{1}{1+u^2}\,du = \frac{1}{3}\tan^{-1}(u) + C$$

Step 4: After evaluating, substitute back in for u:

$$\frac{1}{3}\tan^{-1}(u) + C = \frac{1}{3}\tan^{-1}\left(\frac{x}{3}\right) + C$$

Example 18:

Evaluate $\int \frac{1}{25+9x^2}\,dx$.

Solution:

Since this integral is very similar to one of the inverse trigonometric integration formulas, $\int \frac{1}{1+x^2}\,dx$, first factor out a common perfect square, manipulate the squared term, and then try u-substitution:

$$\int \frac{1}{25+9x^2}\,dx = \int \frac{1}{25\left(1+\frac{9x^2}{25}\right)}\,dx = \frac{1}{25}\int \frac{1}{1+\left(\frac{3x}{5}\right)^2}\,dx$$

Step 1: Let u represent part of the integrand.

Let $u = \frac{3x}{5}$, the input of the squaring function.

Step 2: Differentiate u with respect to x:

$$\frac{d}{dx}(u) = \frac{d}{dx}\left(\frac{3x}{5}\right)$$

$$\frac{du}{dx} = \frac{3}{5}$$

Solve for dx:

$$du = \frac{3}{5} \bullet dx \rightarrow dx = \frac{5}{3}du$$

Step 3: Rewrite the integral in terms of u and evaluate:

$$\frac{1}{25}\int \frac{1}{1+\left(\frac{3x}{5}\right)^2}\,dx \rightarrow \frac{1}{25}\int \frac{1}{1+u^2} \bullet \frac{5}{3}du = \frac{1}{15}\int \frac{1}{1+u^2}\,du$$

$$= \frac{1}{15}\tan^{-1}(u) + C$$

Step 4: After evaluating, substitute back in for u:

$$\frac{1}{15}\tan^{-1}(u) + C = \frac{1}{15}\tan^{-1}\left(\frac{3x}{5}\right) + C$$

Example 19:

Evaluate $\int \frac{5}{\sqrt{16-4x^2}}\,dx$.

Solution:

Since this integral is very similar to one of the inverse trigonometric integration formulas, $\int \frac{1}{\sqrt{1-x^2}}\,dx$, first factor out a common perfect square, manipulate the squared term, and then try u-substitution:

$$\int \frac{5}{\sqrt{16-49x^2}}\,dx = 5\int \frac{1}{\sqrt{16-49x^2}}\,dx = 5\int \frac{1}{\sqrt{16\left(1-\frac{49x^2}{16}\right)}}\,dx$$

$$= 5\int \frac{1}{\sqrt{16}\sqrt{1-\left(\frac{7x}{4}\right)^2}}\,dx = 5\int \frac{1}{4\sqrt{1-\left(\frac{7x}{4}\right)^2}}\,dx$$

$$= \frac{5}{4}\int \frac{1}{\sqrt{1-\left(\frac{7x}{4}\right)^2}}\,dx$$

Step 1: Let u represent part of the integrand.

Let $u = \frac{7x}{4}$, the input of the squaring function.

Step 2: Differentiate u with respect to x:

$$\frac{d}{dx}(u) = \frac{d}{dx}\left(\frac{7x}{4}\right)$$

$$\frac{du}{dx} = \frac{7}{4}$$

Solve for dx:

$$du = \frac{7}{4} \bullet dx \rightarrow dx = \frac{4}{7}du$$

Step 3: Rewrite the integral in terms of u and evaluate:

$$\frac{5}{4}\int \frac{1}{\sqrt{1-\left(\frac{7x}{4}\right)^2}}dx \rightarrow \frac{5}{4}\int \frac{1}{\sqrt{1-u^2}} \bullet \frac{4}{7}du$$

$$= \frac{5}{7}\int \frac{1}{\sqrt{1-u^2}}du = \frac{5}{7}\sin^{-1}(u) + C$$

Step 4: After evaluating, substitute back in for u:

$$\frac{5}{7}\sin^{-1}(u) + C = \frac{5}{7}\sin^{-1}\left(\frac{7x}{4}\right) + C$$

CAUTION—Major Mistake Territory!

The key to recognizing the inverse trigonometric integration formulas is to notice the sum or difference of perfect squares in the denominator of a rational expression with only a constant in the numerator. When manipulating the denominator of the integrand, remember to factor out the constant factor first so there is a 1. Then rewrite the other term so it is a power of 2. The input of this term will be what is used for the *u*-substitution. Which term is manipulated first does matter, which is why in Examples 18 and 19 the constant was factored out first.

BRAIN TICKLERS Set # 25

1. Evaluate $\int \frac{x}{x^2+1}dx$.
2. Evaluate $\int \frac{10}{1+x^2}dx$.
3. Evaluate $\int \frac{40}{x^2+25}dx$.
4. Evaluate $\int \frac{1}{\sqrt{9-16x^2}}dx$.

(Answers are on page 207.)

Integration by Parts

Sometimes when an integrand is a product of two terms, u-substitution cannot be applied because the derivative of the u-term will not match the other term. If this happens, try the technique known as integration by parts.

Integration by parts can be recognized easily when the integrand is a product of two different functions. It follows this formula:

$$\int f(x)g(x)dx = f(x)G(x) - \int f'(x)G(x)dx$$

An easier way to approach integration by parts is to rewrite the above using a different substitution. Let $u = f(x)$ and let $v = G(x) = \int g(x)\,dx$:

$$\int u\,dv = u \bullet v - \int v\,du$$

Integration by parts is *painless*. There are three steps to follow.

Step 1: Select the term that will represent u. Determine the type of functions representing each term in the integrand. Then order them using the helpful mnemonic, LIPET, shown below. The u-term will be whichever function appears first in the mnemonic. The other term will represent dv.

L	I	P	E	T
o	n	o	x	r
g	v	l	p	i
a	e	y	o	g
r	r	n	n	o
i	s	o	e	n
t	e	m	n	o
h		i	t	m
m	T	a	i	e
i	r	l	a	t
c	i		l	r
	g			i
				c

Step 2: Using your let statements, find du by differentiating u and find v by integrating dv, which is finding the antiderivative of dv.

Step 3: Substitute into the formula, $u \bullet v - \int v\,du$, and evaluate the integral. Repeat as necessary.

Example 20:

Evaluate $\int xe^{6x}dx$.

Solution:

Since the integrand is a product of x, and e^{6x}, which are two different functions, integration by parts needs to be used when evaluating.

Step 1: Select the term that will represent u. To do this, use the helpful mnemonic LIPET. The other term will represent dv.

The x represents a polynomial, and e^{6x} represents an exponential function. By LIPET, let $u = x$ and let $dv = e^{6x}dx$.

Step 2: Using your let statements, find du by differentiating u and find v by integrating dv.

If $u = x$, by differentiating with respect to x, $\frac{du}{dx} = 1$ and so $du = dx$.

If $dv = e^{6x}dx$, by integrating with respect to x, $v = \int e^{6x}dx = \frac{e^{6x}}{6}$. (Note: the constant is added at the end of the problem after all integration has been completed.)

Step 3: Substitute into the formula, $u \bullet v - \int v\,du$, and evaluate the integral. Repeat as necessary:

$$\int xe^{6x}dx = x \bullet \frac{e^{6x}}{6} - \int \frac{e^{6x}}{6}dx$$

$$= x \bullet \frac{e^{6x}}{6} - \frac{1}{6}\int e^{6x}dx = \frac{xe^{6x}}{6} - \frac{1}{6} \bullet \frac{e^{6x}}{6} + C$$

$$= \frac{xe^{6x}}{6} - \frac{e^{6x}}{36} + C$$

1+2=3 MATH TALK!

In the previous example, if you tried u-substitution, it would not work because rewriting the integral in terms of u would be impossible. If $u = 6x$, the input of the exponential function, then $\frac{du}{dx} = 6$ or $du = 6dx$. The integrand would still have an x remaining, which did not appear in the derivative. Rewriting the integral would not decrease the amount of terms, and it would not follow one of the integral formulas.

Example 21:

Evaluate $\int x\sin(x)\,dx$.

Solution:

Since the integrand is a product of x and $\sin(x)$, which are two different functions, integration by parts needs to be used when evaluating.

Step 1: Select the term that will represent u. To do this, use the helpful mnemonic LIPET. The other term will represent dv.

The x represents a polynomial, and $\sin(x)$ represents a trigonometric function. By LIPET, let $u = x$ and let $dv = \sin(x)dx$.

Step 2: Using your let statements, find du by differentiating u and find v by integrating dv.

If $u = x$, by differentiating with respect to x, $\frac{du}{dx} = 1$ and so $du = dx$.

If $dv = \sin(x)dx$, by integrating with respect to x, $v = \int \sin(x)\,dx = -\cos(x)$. (Note: the constant is added at the end of the problem after all integration has been completed.)

Step 3: Substitute into the formula, $u \bullet v - \int v\,du$, and evaluate the integral. Repeat as necessary:

$$\int x\sin(x)dx = x \bullet -\cos(x) - \int -\cos(x)dx$$

$$= -x\cos(x) + \int \cos(x)dx = -x\cos(x) + \sin(x) + C$$

Example 22:

Evaluate $\int x^4 \ln(x)\, dx$.

Solution:

Since the integrand is a product of x^4 and $\ln(x)$, which are two different functions, integration by parts needs to be used when evaluating.

Step 1: Select the term that will represent u. To do this, use the helpful mnemonic LIPET. The other term will represent dv.

The x^4 represents a polynomial, and $\ln(x)$ represents a logarithmic function. By LIPET, let $u = \ln(x)$ and let $dv = x^4 dx$.

Step 2: Using your let statements, find du by differentiating u and find v by integrating dv.

If $u = \ln(x)$, by differentiating with respect to x, $\dfrac{du}{dx} = \dfrac{1}{x}$ and so $du = \dfrac{1}{x} dx$.

If $dv = x^4 dx$, by integrating with respect to x, $v = \int x^4 dx = \dfrac{x^5}{5}$. (Note: the constant is added at the end of the problem after all integration has been completed.)

Step 3: Substitute into the formula, $u \bullet v - \int v\, du$, and evaluate the integral. Repeat as necessary:

$$\int x^4 \ln(x) dx = \ln(x) \bullet \frac{x^5}{5} - \int \frac{x^5}{5} \bullet \frac{1}{x} dx$$

$$= \ln(x) \bullet \frac{x^5}{5} - \frac{1}{5} \int x^4 dx = \frac{x^5 \ln(x)}{5} - \frac{1}{5} \bullet \frac{x^5}{5} + C$$

$$= \frac{x^5 \ln(x)}{5} - \frac{x^5}{25} + C$$

Example 23:

Evaluate $\int x^2 \cos(x)\, dx$.

Solution:

Since the integrand is a product of x^2 and $\cos(x)$, which are two different functions, integration by parts needs to be used when evaluating.

Step 1: Select the term that will represent u. To do this, use the helpful mnemonic LIPET. The other term will represent dv.

The x^2 represents a polynomial, and $\cos(x)$ represents a trigonometric function. By LIPET, let $u = x^2$ and let $dv = \cos(x)dx$.

Step 2: Using your let statements, find du by differentiating u and find v by integrating dv.

If $u = x^2$, by differentiating with respect to x, $\frac{du}{dx} = 2x$ and so $du = 2xdx$.

If $dv = \cos(x)dx$, by integrating with respect to x, $v = \int \cos(x)dx = \sin(x)$. (Note: the constant is added at the end of the problem after all integration has been completed.)

Step 3: Substitute into the formula, $u \bullet v - \int v\, du$, and evaluate the integral. Repeat as necessary:

$$\int x^2 \cos(x)dx = x^2 \bullet \sin(x) - \int \sin(x) \bullet 2xdx$$

$$= x^2 \sin(x) - 2\int x \sin(x)dx$$

Integration by parts is necessary again because the new integrand is a product of x and $\sin(x)$, which are two different functions.

Step 1: By LIPET, let $u = x$ and let $dv = \sin(x)dx$.

Step 2: If $u = x$, then $\frac{du}{dx} = 1$ and so $du = dx$.

If $dv = \sin(x)dx$, then $v = \int \sin(x)\, dx = -\cos(x)$. (Note: the constant is added at the end of the problem after all integration has been completed.)

Step 3: Substitute into the formula, $u \bullet v - \int v\,du$, and evaluate the integral. Repeat as necessary:

$$x^2\sin(x) - 2\int x\sin(x)dx$$

$$= x^2\sin(x) - 2\left[x \bullet -\cos(x) - \int -\cos(x)dx\right]$$

$$= x^2\sin(x) - 2[-x\cos(x) + \sin(x)] + C$$

$$= x^2\sin(x) + 2x\cos(x) - 2\sin(x) + C$$

In Example 23, integration by parts was needed twice. Sometimes the functions in the product are cyclic. Unlike polynomial functions, their derivatives do not decrease to a constant but, instead, continue on in a repetitive pattern, like e^x and $\sin(x)$. If this is the case, integration by parts is used until a pattern emerges.

Example 24:

Evaluate $\int e^x \sin(x)\,dx$.

Solution:

Since the integrand is a product of e^x and $\sin(x)$, which are two different functions, integration by parts needs to be used when evaluating.

Step 1: Select the term that will represent u. To do this, use the helpful mnemonic LIPET. The other term will represent dv.

The e^x represents an exponential function, and $\sin(x)$ represents a trigonometric function. By LIPET, let $u = e^x$ and let $dv = \sin(x)dx$.

Step 2: Using your let statements, find du by differentiating u and find v by integrating dv.

If $u = e^x$, by differentiating with respect to x, $\frac{du}{dx} = e^x$ and so $du = e^x dx$.

If $dv = \sin(x)dx$, by integrating with respect to x, $v = \int \sin(x)\,dx = -\cos(x)$. (Note: the constant is added at the end of the problem after all integration has been completed.)

Step 3: Substitute into the formula, $u \cdot v - \int v\,du$, and evaluate the integral. Repeat as necessary:

$$\int e^x \sin(x)dx = e^x \cdot -\cos(x) - \int -\cos(x) \cdot e^x dx$$
$$= -e^x \cos(x) + \int e^x \cos(x)dx$$

Integration by parts is necessary again because the new integrand is a product of e^x and $\cos(x)$, which are two different functions.

Step 1: By LIPET, let $u = e^x$ and let $dv = \cos(x)dx$.

Step 2: If $u = e^x$, then $\frac{du}{dx} = e^x$ and so $du = e^x dx$.
If $dv = \cos(x)dx$, then $v = \int \cos(x)\,dx = \sin(x)$. (Note: the constant is added at the end of the problem after all integration has been completed.)

Step 3: Substitute into the formula, $u \cdot v - \int v\,du$, and evaluate the integral. Repeat as necessary:

$$\underbrace{\int e^x \sin(x)dx} = -e^x \cos(x) + \int e^x \cos(x)dx$$
$$= -e^x \cos(x) + e^x \cdot \sin(x) - \underbrace{\int \sin(x) \cdot e^x dx}$$

A pattern has emerged. On the left side of the equation is the original integral. On the right side of the equation, the same integral has appeared. Much like solving an algebraic equation, the same integrals will be brought to one side of the equation to be isolated:

$$\begin{array}{rl} \int e^x \sin(x)dx = & -e^x \cos(x) + e^x \sin(x) - \int e^x \sin(x)dx \\ +\int e^x \sin(x)dx & \qquad\qquad\qquad\qquad +\int e^x \sin(x)dx \\ \hline \end{array}$$

$$2\int e^x \sin(x)dx = -e^x \cos(x) + e^x \sin(x)$$

Divide both sides by 2:

$$\int e^x \sin(x)dx = \frac{-e^x \cos(x)}{2} + \frac{e^x \sin(x)}{2}$$

The final answer will include the constant that is from the integration in the previous steps:

$$\int e^x \sin(x)dx = \frac{-e^x \cos(x)}{2} + \frac{e^x \sin(x)}{2} + C$$

Integration by parts can also be applied when asked to integrate a single nonpolynomial function that is not on the list of integration formulas.

Example 25:

Evaluate $\int \ln(x)\,dx$.

Solution:

Since the integrand is a single nonpolynomial function not on the list of integration formulas, integration by parts needs to be used when evaluating.

To think of it as a product of two terms, write the integral as $\int \ln(x) \bullet 1dx$.

Step 1: Select the term that will represent u. To do this, use the helpful mnemonic LIPET. The other term will represent dv.

The $\ln(x)$ represents a logarithmic function, and 1 represents a polynomial. By LIPET, let $u = \ln(x)$ and let $dv = 1dx$.

Step 2: Using your let statements, find du by differentiating u and find v by integrating dv.

If $u = \ln(x)$, by differentiating with respect to x, $\frac{du}{dx} = \frac{1}{x}$ and so $du = \frac{1}{x}dx$.

If $dv = 1dx$, by integrating with respect to x, $v = \int 1\,dx = x$. (Note: the constant is added at the end of the problem after all integration has been completed.)

Step 3: Substitute into the formula, $u \bullet v - \int v\,du$, and evaluate the integral. Repeat as necessary:

$$\int \ln(x)dx = \ln(x) \bullet x - \int x \bullet \frac{1}{x}dx$$

$$= \ln(x) \bullet x - \int 1dx = x\ln(x) - x + C$$

Example 26:

Evaluate $\int \tan^{-1}(x)\,dx$.

Solution:

Since the integrand is a single nonpolynomial function not on the list of integration formulas, integration by parts is necessary when evaluating.

To think of it as a product of two terms, write the integral as $\int \tan^{-1}(x) \bullet 1dx$.

Step 1: Select the term that will represent u. To do this, use the helpful mnemonic LIPET. The other term will represent dv.

The $\tan^{-1}(x)$ represents an inverse trigonometric function, and 1 represents a polynomial. By LIPET, let $u = \tan^{-1}(x)$ and let $dv = 1dx$.

Step 2: Using your let statements, find du by differentiating u and find v by integrating dv.

If $u = \tan^{-1}(x)$, by differentiating with respect to x, $\frac{du}{dx} = \frac{1}{1+x^2}$ and so $du = \frac{1}{1+x^2}dx$.

If $dv = 1dx$, by integrating with respect to x, $v = \int 1\,dx = x$. (Note: the constant is added at the end of the problem after all integration has been completed.)

Step 3: Substitute into the formula, $u \bullet v - \int v\,du$, and evaluate the integral. Repeat as necessary:

$$\int \tan^{-1}(x)\,dx = \tan^{-1}(x) \bullet x - \int x \bullet \frac{1}{1+x^2}dx$$

To evaluate the integral, $\int x \bullet \frac{1}{1+x^2}dx$, use u-substitution. Let $u = 1 + x^2$, the denominator of the integrand. Differentiating gives $\frac{du}{dx} = 2x$. The derivative has x, which is the other term in the integrand. We can make a substitution for this instead. Solving for xdx gives $du = 2xdx$, and so $xdx = \frac{1}{2}du$.

Substitute into the integral:

$$\int x \bullet \frac{1}{1+x^2}dx = \int \frac{1}{1+x^2} \bullet xdx \to \int \frac{1}{u} \bullet \frac{1}{2}du$$

$$= \frac{1}{2}\ln|u| + C = \frac{1}{2}\ln\left|1+x^2\right| + C$$

Substitute back into the integral:

$$\int \tan^{-1}(x)dx = \tan^{-1}(x) \bullet x - \int x \bullet \frac{1}{1+x^2}dx$$

$$\int \tan^{-1}(x)dx = x\tan^{-1}(x) - \frac{1}{2}\ln\left|1+x^2\right| + C$$

BRAIN TICKLERS Set # 26

1. Evaluate $\int xe^x dx$.
2. Evaluate $\int 2t\cos(3t)\,dt$.
3. Evaluate $\int e^{-x}\cos(x)\,dx$.
4. Evaluate $\int \ln(x^3)\,dx$.

(Answers are on page 207.)

Integration by Partial Fraction Decomposition

Sometimes when an integrand is a quotient of two terms, u-substitution cannot be applied because the derivative of the u-term will not match the other term. If this happens, try the technique known as *partial fraction decomposition*.

If the integrand is a rational function, $\frac{f(x)}{g(x)}$, where $f(x)$ is a lower degree than $g(x)$ and where $g(x)$ is easily factorable, try integrating by partial fraction decomposition. This book will focus on this technique only when the denominator can be written as a product of nonrepeating linear factors.

This technique breaks down a rational function into the sum or difference of smaller expressions that can be integrated more easily. For example, $\frac{1}{x-3}+\frac{1}{x+2}=\frac{x+2+x-3}{(x-3)(x+2)}=\frac{2x-1}{x^2-x-6}$. So, if asked to evaluate $\int \frac{2x-1}{x^2-x-6}\,dx$, it would be easier to integrate its equivalent sum, $\int\left(\frac{1}{x-3}+\frac{1}{x+2}\right)dx$, instead of the original expression since the smaller rational expressions can be integrated separately. Partial fraction decomposition gets its name since it involves taking the rational integrand and "decomposing" or breaking it down into an equivalent sum or difference of smaller rational functions.

Partial fraction decomposition is *painless*. There are four steps to follow.

Step 1: Factor the denominator of the integrand into a product of nonrepeating linear factors.

Step 2: Each factor is the denominator of a rational expression with an unknown constant as its numerator.

Step 3: Set the sum of the rational expressions equal to the integrand, and solve for the unknown constants.

Step 4: Substitute the values of each constant into the sum of rational functions, and integrate each new rational expression.

Example 27:

Evaluate $\int \frac{3}{x^2+x-2}\,dx$.

Solution:

Since the integrand is a rational function whose numerator is a lower degree than the denominator and since the denominator is factorable, try partial fraction decomposition.

Step 1: Factor the denominator of the integrand into a product of nonrepeating linear factors:

$$x^2 + x - 2 = (x + 2)(x - 1)$$

Step 2: Each factor is the denominator of a rational expression with an unknown constant as its numerator:

$$\frac{A}{x+2} + \frac{B}{x-1}$$

Step 3: Set the sum of the rational expressions equal to the integrand, and solve for the unknown constants:

$$\frac{3}{x^2 + x - 2} = \frac{A}{x+2} + \frac{B}{x-1}$$

To solve for A and B, multiply the equation by the factors $(x + 2)(x - 1)$ to reduce the denominators:

$$(x+2)(x-1)\left[\frac{3}{x^2 + x - 2}\right] = (x+2)(x-1)\left[\frac{A}{x+2} + \frac{B}{x-1}\right]$$

$$3 = A(x - 1) + B(x + 2)$$

To solve for A and B, substitute different values of x that will zero out each factor.

Let $x = 1$	Let $x = -2$
$3 = A((1) - 1) + B((1) + 2)$	$3 = A((-2) - 1) + B((-2) + 2)$
$3 = A(0) + B(3)$	$3 = A(-3) + B(0)$
$3 = 3B$	$3 = -3A$
$B = 1$	$A = -1$

Step 4: Substitute the values of each constant into the sum of rational functions, and integrate each new rational expression:

$$\frac{3}{x^2 + x - 2} = \frac{-1}{x+2} + \frac{1}{x-1}$$

$$\int \frac{-1}{x+2}\,dx + \int \frac{1}{x-1}\,dx$$

These integrals can be evaluated using u-substitution.

$$\int \frac{-1}{x+2}dx$$

Let $u = x + 2$

$$\frac{du}{dx} = 1 \rightarrow du = 1dx$$

$$\int \frac{-1}{x+2}dx \rightarrow -\int \frac{1}{u}du$$
$$= -\ln|u| + C$$
$$= -\ln|x+2| + C$$

$$\int \frac{1}{x-1}dx$$

Let $u = x - 1$

$$\frac{du}{dx} = 1 \rightarrow du = 1dx$$

$$\int \frac{1}{x-1}dx \rightarrow \int \frac{1}{u}du = \ln|u| + C$$
$$= \ln|x-1| + C$$

$$\int \frac{3}{x^2+x-2}dx = \int \frac{-1}{x+2}dx + \int \frac{1}{x-1}dx$$
$$= -\ln|x+2| + \ln|x-1| + C$$

1+2=3 MATH TALK!

The constants from each separate integral are combined at the end of the evaluation as one general constant or one "+ C" and not written as a combination of like terms, such as "+ 2C."

Example 28:

Evaluate $\int \frac{x+7}{x^2-x-6}dx$.

Solution:

Since the integrand is a rational function whose numerator is a lower degree than the denominator and since the denominator is factorable, try partial fraction decomposition.

Step 1: Factor the denominator of the integrand into a product of nonrepeating linear factors:

$$x^2 - x - 6 = (x+2)(x-3)$$

Step 2: Each factor is the denominator of a rational expression with an unknown constant as its numerator:

$$\frac{A}{x+2}+\frac{B}{x-3}$$

Step 3: Set the sum of the rational expressions equal to the integrand, and solve for the unknown constants:

$$\frac{x+7}{x^2-x-6}=\frac{A}{x+2}+\frac{B}{x-3}$$

To solve for A and B, multiply the equation by the factors $(x+2)(x-3)$ to reduce the denominators:

$$(x+2)(x-3)\left[\frac{x+7}{x^2-x-6}\right]=(x+2)(x-3)\left[\frac{A}{x+2}+\frac{B}{x-3}\right]$$

$$x+7=A(x-3)+B(x+2)$$

To solve for A and B, substitute different values of x that will zero out each factor.

Let $x = 3$	Let $x = -2$
$(3)+7=A((3)-3)+B((3)+2)$	$(-2)+7=A((-2)-3)+B((-2)+2)$
$10=A(0)+B(5)$	$5=A(-5)+B(0)$
$10=5B$	$5=-5A$
$B=2$	$A=-1$

Step 4: Substitute the values of each constant into the sum of rational functions, and integrate each new rational expression:

$$\frac{x+7}{x^2-x-6}=\frac{-1}{x+2}+\frac{2}{x-3}$$

$$\int\frac{-1}{x+2}dx+\int\frac{2}{x-3}dx$$

These integrals can be evaluated using u-substitution.

$$\int \frac{-1}{x+2}dx$$

Let $u = x + 2$

$$\frac{du}{dx} = 1 \to du = 1dx$$

$$\int \frac{-1}{x+2}dx \to -\int \frac{1}{u}du$$
$$= -\ln|u| + C = -\ln|x+2| + C$$

$$\int \frac{2}{x-3}dx$$

Let $u = x - 3$

$$\frac{du}{dx} = 1 \to du = 1dx$$

$$\int \frac{2}{x-3}dx \to 2\int \frac{1}{u}du$$
$$= 2\ln|u| + C = 2\ln|x-3| + C$$

$$\int \frac{x+7}{x^2 - x - 6}dx = \int \frac{-1}{x+2}dx + \int \frac{2}{x-3}dx$$
$$= -\ln|x+2| + 2\ln|x-3| + C$$

Example 29:

Evaluate $\int \frac{11x+17}{2x^2+7x-4}dx$.

Solution:

Since the integrand is a rational function whose numerator is a lower degree than the denominator and since the denominator is factorable, try partial fraction decomposition.

Step 1: Factor the denominator of the integrand into a product of nonrepeating linear factors:

$$2x^2 + 7x - 4 = (2x-1)(x+4)$$

Step 2: Each factor is the denominator of a rational expression with an unknown constant as its numerator:

$$\frac{A}{2x-1} + \frac{B}{x+4}$$

Step 3: Set the sum of the rational expressions equal to the integrand, and solve for the unknown constants:

$$\frac{11x+17}{2x^2+7x-4} = \frac{A}{2x-1} + \frac{B}{x+4}$$

To solve for A and B, multiply the equation by the factors $(2x-1)(x+4)$ to reduce the denominators:

$$(2x-1)(x+4)\left[\frac{11x+17}{2x^2+7x-4}\right]=(2x-1)(x+4)\left[\frac{A}{2x-1}+\frac{B}{x+4}\right]$$

$$11x+17=A(x+4)+B(2x-1)$$

To solve for A and B, substitute different values of x that will zero out each factor.

Let $x=-4$

$$11(-4)+17=A((-4)+4)+B(2(-4)-1)$$
$$-27=A(0)+B(-9)$$
$$-27=-9B$$
$$B=3$$

Let $x=\frac{1}{2}$

$$11\left(\frac{1}{2}\right)+17=A\left(\left(\frac{1}{2}\right)+4\right)+B\left(2\left(\frac{1}{2}\right)-1\right)$$
$$22.5=A(4.5)+B(0)$$
$$22.5=4.5A$$
$$A=5$$

Step 4: Substitute the values of each constant into the sum of rational functions, and integrate each new rational expression:

$$\frac{11x+17}{2x^2+7x-4}=\frac{5}{2x-1}+\frac{3}{x+4}$$

$$\int\frac{5}{2x-1}dx+\int\frac{3}{x+4}dx$$

These integrals can be evaluated using u-substitution.

$$\int\frac{5}{2x-1}dx$$

Let $u=2x-1$

$$\frac{du}{dx}=2\rightarrow du=2dx\rightarrow dx=\frac{1}{2}du$$

$$\int\frac{5}{2x-1}dx\rightarrow 5\int\frac{1}{u}\bullet\frac{1}{2}du=\frac{5}{2}\ln|u|+C$$

$$=\frac{5}{2}\ln|2x-1|+C$$

$$\int\frac{3}{x+4}dx$$

Let $u=x+4$

$$\frac{du}{dx}=1\rightarrow du=1dx$$

$$\int\frac{3}{x+4}dx\rightarrow 3\int\frac{1}{u}du=3\ln|u|+C$$

$$=3\ln|x+4|+C$$

$$\int\frac{11x+17}{2x^2+7x-4}dx=\int\frac{5}{2x-1}dx+\int\frac{3}{x+4}dx$$

$$=\frac{5}{2}\ln|2x-1|+3\ln|x+4|+C$$

Example 30:

Evaluate $\int \frac{dx}{(x-5)(x-2)}$.

Solution:

Since the integrand is a rational function whose numerator is a lower degree than the denominator and since the denominator is already factored, try partial fraction decomposition.

Step 1: The denominator of the integrand is already factored into a product of nonrepeating linear factors:

$$(x-5)(x-2)$$

Step 2: Each factor is the denominator of a rational expression with an unknown constant as its numerator:

$$\frac{A}{x-5} + \frac{B}{x-2}$$

Step 3: Set the sum of the rational expressions equal to the integrand, and solve for the unknown constants:

$$\frac{1}{(x-5)(x-2)} = \frac{A}{x-5} + \frac{B}{x-2}$$

To solve for A and B, multiply the equation by the factors $(x-5)(x-2)$ to reduce the denominators:

$$(x-5)(x-2)\left[\frac{1}{(x-5)(x-2)}\right] = (x-5)(x-2)\left[\frac{A}{x-5} + \frac{B}{x-2}\right]$$

$$1 = A(x-2) + B(x-5)$$

To solve for A and B, substitute different values of x that will zero out each factor.

Let $x = 2$	Let $x = 5$
$1 = A((2)-2) + B((2)-5)$	$1 = A((5)-2) + B((5)-5)$
$1 = A(0) + B(-3)$	$1 = A(3) + B(0)$
$1 = -3B$	$1 = 3A$
$B = -\frac{1}{3}$	$A = \frac{1}{3}$

Step 4: Substitute the values of each constant into the sum of rational functions, and integrate each new rational expression:

$$\frac{1}{(x-5)(x-2)} = \frac{\frac{1}{3}}{x-5} + \frac{-\frac{1}{3}}{x-2}$$

$$\int \frac{\frac{1}{3}}{x-5}dx + \int \frac{-\frac{1}{3}}{x-2}dx$$

These integrals can be evaluated using u-substitution.

$$\int \frac{\frac{1}{3}}{x-5}dx$$

Let $u = x - 5$

$$\frac{du}{dx} = 1 \rightarrow du = 1dx$$

$$\int \frac{\frac{1}{3}}{x-5}dx \rightarrow \frac{1}{3}\int \frac{1}{u}du$$

$$= \frac{1}{3}\ln|u| + C = \frac{1}{3}\ln|x-5| + C$$

$$\int \frac{-\frac{1}{3}}{x-2}dx$$

Let $u = x - 2$

$$\frac{du}{dx} = 1 \rightarrow du = 1dx$$

$$\int \frac{-\frac{1}{3}}{x-2}dx \rightarrow -\frac{1}{3}\int \frac{1}{u}du$$

$$= -\frac{1}{3}\ln|u| + C = -\frac{1}{3}\ln|x-2| + C$$

$$\int \frac{1}{(x-5)(x-2)}dx = \int \frac{\frac{1}{3}}{x-5}dx + \int \frac{-\frac{1}{3}}{x-2}dx$$

$$= \frac{1}{3}\ln|x-5| - \frac{1}{3}\ln|x-2| + C$$

BRAIN TICKLERS Set # 27

1. Evaluate $\int \frac{dx}{(x+3)(x-1)}$.
2. Evaluate $\int \frac{1}{2x^2+x}\,dx$.
3. Evaluate $\int \frac{x}{x^2-5x+6}\,dx$.
4. Evaluate $\int \frac{5-x}{2x^2+x-1}\,dx$.

(Answers are on page 207.)

Integration by Completing the Square

When asked to evaluate a rational function, where u-substitution does not apply and where the denominator is not factorable, try rewriting the denominator by completing the square. Often this will rewrite the rational function in a form that matches an inverse trigonometric integration formula.

Rewriting a quadratic, $ax^2 + bx + c$, into a perfect squares trinomial by completing the square involves four steps.

Step 1: Have the leading term, a, equal to 1. If not, factor.

Step 2: Add and subtract $\left(\frac{b}{2}\right)^2$ to the expression.

Step 3: Simplify and then factor the first three terms; this will create the perfect square.

Step 4: Combine any remaining like terms.

Example 31:

Evaluate $\int \frac{dx}{x^2-2x+17}$.

Solution:

The technique of u-substitution does not apply here since the only expression is in the denominator and its derivative would not substitute anywhere else in the integrand. The denominator is also not factorable. Rewrite the denominator as a perfect squares trinomial by completing the square.

Step 1: Have the leading term, a, equal to 1. If not, factor.

The leading term is 1.

Step 2: Add and subtract $\left(\frac{b}{2}\right)^2$ to the expression:

$$x^2 - 2x + \left(\frac{-2}{2}\right)^2 - \left(\frac{-2}{2}\right)^2 + 17$$

Step 3: Simplify and then factor the first three terms; this will create the perfect square:

$$x^2 - 2x + 1 - 1 + 17$$

$$(x-1)^2 - 1 + 17$$

$$(x-1)^2 + 16$$

Step 4: Combine any remaining like terms:

$$\int \frac{dx}{x^2 - 2x + 17} = \int \frac{1}{(x-1)^2 + 16}\,dx$$

Step 5: This resembles the inverse trigonometric integration formula $\int \frac{1}{1+x^2}\,dx = \tan^{-1}(x) + C$. Factor out the perfect square, and use u-substitution to evaluate:

$$\int \frac{1}{16\left(\frac{(x-1)^2}{16} + 1\right)}\,dx = \frac{1}{16}\int \frac{1}{\left(\frac{x-1}{4}\right)^2 + 1}\,dx$$

$$\text{Let } u = \frac{x-1}{4}$$

$$\frac{du}{dx} = \frac{1}{4} \rightarrow dx = 4du$$

$$\frac{1}{16}\int \frac{1}{\left(\frac{x-1}{4}\right)^2 + 1}\,dx \rightarrow \frac{1}{16}\int \frac{1}{u^2 + 1} \bullet 4du = \frac{1}{4}\int \frac{1}{1+u^2}\,du$$

$$= \frac{1}{4}\tan^{-1}(u) + C$$

$$\int \frac{dx}{x^2 - 2x + 17} = \frac{1}{4}\tan^{-1}\left(\frac{x-1}{4}\right) + C$$

Example 32:

Evaluate $\int \frac{dx}{\sqrt{4x - x^2}}$.

Solution:

The technique of u-substitution does not apply here since the only expression is in the denominator and its derivative would not substitute anywhere else in the integrand. Rewrite the denominator as a perfect squares trinomial by completing the square.

Step 1: Have the leading term, a, equal to 1. If not, factor. Factor out a -1 from each term:

$$-(x^2 - 4x)$$

Step 2: Add and subtract $\left(\frac{b}{2}\right)^2$ to the expression:

$$-\left(x^2 - 4x + \left(\frac{-4}{2}\right)^2 - \left(\frac{-4}{2}\right)^2\right)$$

Step 3: Simplify and then factor the first three terms; this will create the perfect square.

Simplify:

$$-(x^2 - 4x + 4 - 4)$$

Factor the first three terms in the parentheses:

$$-((x - 2)^2 - 4)$$

Step 4: Distribute the -1 to the two terms:

$$-(x - 2)^2 + 4$$

$$4 - (x - 2)^2$$

$$\int \frac{dx}{\sqrt{4x - x^2}} = \int \frac{1}{\sqrt{4 - (x - 2)^2}}\,dx$$

Step 5: This resembles the inverse trigonometric integration formula: $\int \frac{1}{\sqrt{1-x^2}}dx = \sin^{-1}(x) + C$. Factor out the perfect square, and use u-substitution to evaluate:

$$\int \frac{1}{\sqrt{4\left(1-\frac{(x-2)^2}{4}\right)}}dx = \int \frac{1}{\sqrt{4}\sqrt{1-\frac{(x-2)^2}{4}}}dx$$

$$= \int \frac{1}{2\sqrt{1-\frac{(x-2)^2}{4}}}dx = \frac{1}{2}\int \frac{1}{\sqrt{1-\left(\frac{x-2}{2}\right)^2}}dx$$

$$\text{Let } u = \frac{x-2}{2}$$

$$\frac{du}{dx} = \frac{1}{2} \rightarrow dx = 2du$$

$$\frac{1}{2}\int \frac{1}{\sqrt{1-\left(\frac{x-2}{2}\right)^2}}dx \rightarrow \frac{1}{2}\int \frac{1}{\sqrt{1-u^2}} \bullet 2du = \int \frac{1}{\sqrt{1-u^2}}du$$

$$= \sin^{-1}(u) + C$$

$$\int \frac{dx}{\sqrt{4x-x^2}} = \sin^{-1}\left(\frac{x-2}{2}\right) + C$$

Integration with Powers of Trigonometric Functions

Integrands with trigonometric functions raised to powers must be rewritten to match the trigonometric integration formulas. To do this, trigonometric identities must be substituted into the integrands.

The table below lists important trigonometric identities that can be used as substitutions.

Pythagorean Trigonometric Identities	Double Angle Trigonometric Identities
$\sin^2(\theta) = 1 - \cos^2(\theta)$	$\sin^2(\theta) = \cos^2(\theta) - \cos(2\theta)$
$\cos^2(\theta) = 1 - \sin^2(\theta)$	$\sin^2(\theta) = \frac{1 - \cos(2\theta)}{2}$
$\tan^2(\theta) = \sec^2(\theta) - 1$	$\cos^2(\theta) = \sin^2(\theta) + \cos(2\theta)$
$\sec^2(\theta) = \tan^2(\theta) + 1$	$\cos^2(\theta) = \frac{1 + \cos(2\theta)}{2}$
$\csc^2(\theta) = 1 + \cot^2(\theta)$	
$\cot^2(\theta) = \csc^2(\theta) - 1$	

Example 33:

Evaluate $\int \cos^3(x)\sin^4(x)dx$.

Solution:

Having an integrand with cos(x) and sin(x) is perfect for u-substitution since there is a relationship between the two functions and their derivatives. The challenge is the powers. Rewrite the integrand so that a Pythagorean trigonometric identity can be substituted for cos(x):

$$\int \cos^3(x)\sin^4(x)dx = \int \cos^2(x) \bullet \cos(x) \bullet \sin^4(x)dx$$
$$= \int (1 - \sin^2(x))\cos(x)\sin^4(x)dx$$

Distribute into the parentheses, and write as two separate integrals:

$$\int (1 - \sin^2(x))\cos(x)\sin^4(x)dx = \int (\sin^4(x)\cos(x) - \sin^6(x)\cos(x))dx$$
$$= \int \sin^4(x)\cos(x)dx - \int \sin^6(x)\cos(x)dx$$

Now u-substitution can be used for each integral.

$$\int \sin^4(x)\cos(x)dx$$

Let $u = \sin(x)$

$$\frac{du}{dx} = \cos(x) \rightarrow du = \cos(x)dx$$

$$\int \sin^4(x)\cos(x)dx \rightarrow \int u^4\,du = \frac{u^5}{5} + C$$

$$\int \sin^4(x)\cos(x)dx = \frac{\sin^5(x)}{5} + C$$

$$\int \sin^6(x)\cos(x)dx$$

Let $u = \sin(x)$

$$\frac{du}{dx} = \cos(x) \rightarrow du = \cos(x)dx$$

$$\int \sin^6(x)\cos(x)dx \rightarrow \int u^6 du = \frac{u^7}{7} + C$$

$$\int \sin^6(x)\cos(x)dx = \frac{\sin^7(x)}{7} + C$$

$$\int \cos^3(x)\sin^4(x)dx = \frac{\sin^5(x)}{5} - \frac{\sin^7(x)}{7} + C$$

Example 34:

Evaluate $\int \sin^4(x)dx$.

Solution:

This is a trigonometric function raised to a power. Rewrite the integrand so that a double angle trigonometric identity can be substituted in for $\sin(x)$:

$$\int (\sin^2(x))^2\,dx = \int \left(\frac{1-\cos(2x)}{2}\right)^2 dx$$

Expand the integrand:

$$\int \left(\frac{1-\cos(2x)}{2}\right)^2 dx = \int \frac{1-2\cos(2x)+\cos^2(2x)}{4}dx$$

$$= \frac{1}{4}\int 1 - 2\cos(2x) + \cos^2(2x)dx$$

Substitute a double angle trigonometric identify for the power of $\cos(x)$:

$$\frac{1}{4}\int 1 - 2\cos(2x) + \cos^2(2x)dx = \frac{1}{4}\int 1 - 2\cos(2x) + \frac{1+\cos(4x)}{2}dx$$

Separate the integrals:

$$\frac{1}{4}\int 1 - 2\cos(2x) + \frac{1+\cos(4x)}{2}dx$$
$$= \frac{1}{4}\left[\int 1dx - 2\int \cos(2x)dx + \frac{1}{2}\int 1dx + \frac{1}{2}\int \cos(4x)dx\right]$$

Now u-substitution can be used for the two trigonometric integrals.

$$2\int \cos(2x)dx$$

Let $u = 2x$

$$\frac{du}{dx} = 2 \rightarrow dx = \frac{1}{2}du$$

$$2\int \cos(2x)dx \rightarrow 2\int \cos(u) \cdot \frac{1}{2}du$$

$$= 2 \cdot \frac{1}{2}\int \cos(u)du = \sin(u) + C$$

$$2\int \cos(2x)dx = \sin(2x) + C$$

$$\frac{1}{2}\int \cos(4x)dx$$

Let $u = 4x$

$$\frac{du}{dx} = 4 \rightarrow dx = \frac{1}{4}du$$

$$\frac{1}{2}\int \cos(4x)dx \rightarrow \frac{1}{2}\int \cos(u) \cdot \frac{1}{4}du$$

$$= \frac{1}{2} \cdot \frac{1}{4}\sin(u) + C$$

$$\frac{1}{2}\int \cos(4x)dx = \frac{1}{8}\sin(4x) + C$$

$$\int \sin^4(x)\,dx = \frac{1}{4}\left[\int 1dx - 2\int \cos(2x)\,dx + \frac{1}{2}\int 1dx + \frac{1}{2}\int \cos(4x)dx\right]$$
$$= \frac{1}{4}\left[x - \sin(2x) + \frac{1}{2}x + \frac{1}{8}\sin(4x)\right] + C$$

Distribute and combine like terms:

$$\int \sin^4(x)dx = \frac{1}{4}x - \frac{1}{4}\sin(2x) + \frac{1}{8}x + \frac{1}{32}\sin(4x) + C$$

$$\int \sin^4(x)dx = \frac{3}{8}x - \frac{1}{4}\sin(2x) + \frac{1}{32}\sin(4x) + C$$

Integration by Trigonometric Substitutions

If an integrand contains radicals of the form $\sqrt{a^2 - x^2}$, $\sqrt{x^2 + a^2}$, or $\sqrt{x^2 - a^2}$, use integration by trigonometric substitutions. The goal is to make a substitution for the x-term that will use trigonometric identities to help eliminate the radical.

In general if $\sqrt{a^2 - x^2}$, let $x = a\sin(\theta)$:

$$\sqrt{a^2 - x^2} \to \sqrt{a^2 - (a\sin(\theta))^2} = \sqrt{a^2 - a^2\sin^2(\theta)}$$
$$= \sqrt{a^2(1 - \sin^2(\theta))}$$

By the Pythagorean trigonometric identity, $\sin^2\theta + \cos^2\theta = 1$:

$$\sqrt{a^2(1 - \sin^2(\theta))} = \sqrt{a^2\cos^2(\theta)} = a\cos(\theta)$$

Making this substitution rewrites the expression to make it easier to integrate.

The table below summarizes the different trigonometric substitutions to use for integration.

Integrand Expression	Substitution	Restriction on θ	Simplification Under Radical
$\sqrt{a^2 - x^2}$	$x = a\sin(\theta)$	$-\frac{\pi}{2} \leq \theta \leq \frac{\pi}{2}$	$a^2 - x^2 = a^2 - a^2\sin^2(\theta)$ $= a^2(1 - \sin^2(\theta))$ $= a^2\cos^2(\theta)$
$\sqrt{a^2 + x^2}$	$x = a\tan(\theta)$	$-\frac{\pi}{2} \leq \theta \leq \frac{\pi}{2}$	$a^2 + x^2 = a^2 + a^2\tan^2(\theta)$ $= a^2(1 + \tan^2(\theta))$ $= a^2\sec^2(\theta)$
$\sqrt{x^2 - a^2}$	$x = a\sec(\theta)$	If $x \geq a$, $0 \leq \theta \leq \frac{\pi}{2}$ If $x \leq -a$, $\frac{\pi}{2} \leq \theta \leq \pi$	$x^2 - a^2 = a^2\sec^2(\theta) - a^2$ $= a^2(\sec^2(\theta) - 1)$ $= a^2\tan^2(\theta)$

Example 35:

Evaluate $\int \frac{dx}{x^2\sqrt{4-x^2}}$.

Solution:

The technique of *u*-substitution does not apply here since the only expression is in the denominator and its derivative would not substitute anywhere else in the integrand. Since the denominator does not match one of the inverse trigonometric integration formulas, do not use completing the square. Notice that the integrand contains a radical of the form $\sqrt{a^2 - x^2}$, where $a = 2$.

Using the table, let $x = 2\sin(\theta)$. This means that $x^2 = 4\sin^2(\theta)$.

Differentiate with respect to θ:

$$\frac{dx}{d\theta} = 2\cos(\theta) \rightarrow dx = 2\cos(\theta)d\theta$$

Substitute into the integral:

$$\int \frac{dx}{x^2\sqrt{4-x^2}} \rightarrow \int \frac{2\cos(\theta)d\theta}{4\sin^2(\theta)\sqrt{4-4\sin^2(\theta)}}$$

Simplify the radical in the denominator:

$$\int \frac{2\cos(\theta)d\theta}{4\sin^2(\theta)\sqrt{4-4\sin^2(\theta)}} = \int \frac{2\cos(\theta)d\theta}{4\sin^2(\theta)\sqrt{4(1-\sin^2(\theta))}}$$

$$= \int \frac{2\cos(\theta)d\theta}{4\sin^2(\theta)\sqrt{4\cos^2\theta}} = \int \frac{2\cos(\theta)d\theta}{4\sin^2(\theta)\bullet 2\cos(\theta)}$$

Reduce the common factor in the numerator and denominator:

$$\int \frac{2\cos(\theta)d\theta}{4\sin^2(\theta)\bullet 2\cos(\theta)} = \int \frac{d\theta}{4\sin^2(\theta)} = \frac{1}{4}\int \frac{d\theta}{\sin^2(\theta)}$$

$$= \frac{1}{4}\int \csc^2(\theta)d\theta = -\frac{1}{4}\cot(\theta) + C$$

The solution is in terms of θ, but the original integral was in terms of x. To rewrite the solution in terms of x, use right triangle trigonometry.

Since $x = 2\sin(\theta)$, then $\sin(\theta) = \frac{x}{2}$.

This ratio can be used to fill in two sides of a right triangle since $\sin(\theta) = \frac{\text{opposite}}{\text{hypotenuse}}$. When drawing a right triangle, the side opposite to angle θ will be x and the hypotenuse will be 2.

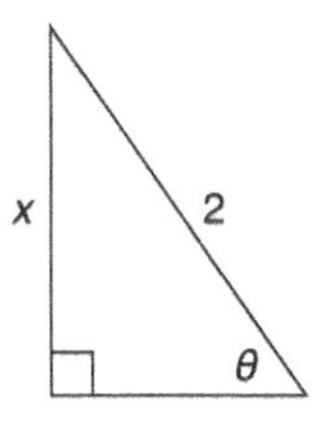

By using the Pythagorean Theorem, the third side of the triangle can be found in terms of x.

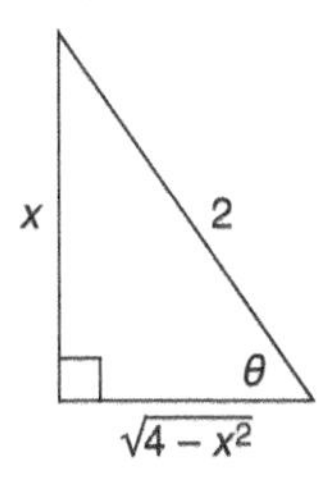

Third side $= \sqrt{2^2 - x^2} = \sqrt{4 - x^2}$.

Using the sides of the right triangle, find $\cot(\theta)$:

$$\cot(\theta) = \frac{\text{adjacent}}{\text{opposite}} = \frac{\sqrt{4 - x^2}}{x}$$

Substitute into the solution:

$$\int \frac{dx}{x^2\sqrt{4 - x^2}} = -\frac{1}{4}\cot(\theta) + C = -\frac{1}{4} \bullet \frac{\sqrt{4 - x^2}}{x} + C$$

Example 36:

Evaluate $\int \frac{\sqrt{x^2 - 25}}{x} dx$.

Solution:

Since the integrand contains a radical of the form $\sqrt{x^2 - 25}$, where $a = 5$, use integration by trigonometric substitutions.

Using the table, let $x = 5\sec(\theta)$. This means that $x^2 = 25\sec^2(\theta)$.

Differentiate with respect to θ.

$$\frac{dx}{d\theta} = 5\tan(\theta)\sec(\theta) \rightarrow dx = 5\tan(\theta)\sec(\theta)d\theta$$

Substituting into the integral:

$$\int \frac{\sqrt{x^2 - 25}}{x} dx \rightarrow \int \frac{\sqrt{25\sec^2(\theta) - 25}}{5\sec(\theta)} \bullet 5\tan(\theta)\sec(\theta)d\theta$$

Use the table to simplify under the radical:

$$\int \frac{\sqrt{25\sec^2(\theta) - 25}}{5\sec(\theta)} \bullet 5\tan(\theta)\sec(\theta)d\theta$$

$$= \int \frac{\sqrt{25\tan^2(\theta)}}{5\sec(\theta)} \bullet 5\tan(\theta)\sec(\theta)d\theta$$

Simplify the integrand:

$$\int \sqrt{25\tan^2(\theta)} \bullet \tan(\theta)d\theta = \int 5\tan^2(\theta)d\theta$$

Substitute a Pythagorean trigonometric identity:

$$\int 5\tan^2(\theta)d\theta = 5\int (\sec^2(\theta) - 1)d\theta$$

Separate and evaluate each integral:

$$5\int (\sec^2(\theta) - 1)d\theta = 5\int \sec^2(\theta)d\theta - 5\int 1d\theta = 5\tan(\theta) - 5\theta + C$$

The solution is in terms of θ, but the original integral was in terms of x. To rewrite the solution in terms of x, use right triangle trigonometry.

Since $x = 5\sec(\theta)$, then $\sec(\theta) = \frac{x}{5}$.

To solve for θ, rewrite using inverse trigonometry:

$$\theta = \sec^{-1}\left(\frac{x}{5}\right)$$

The secant ratio can also be used to fill in two sides of a right triangle since $\sec(\theta) = \frac{\text{hypotenuse}}{\text{adjacent}}$. When drawing a right triangle, the side adjacent to angle θ will be 5 and the hypotenuse will be x.

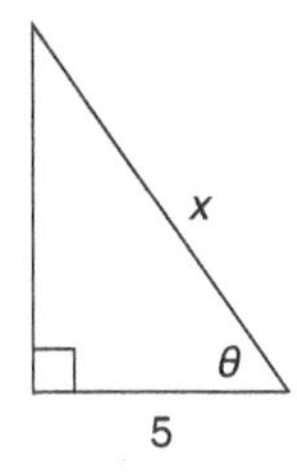

By using the Pythagorean Theorem, the third side of the triangle can be found in terms of x.

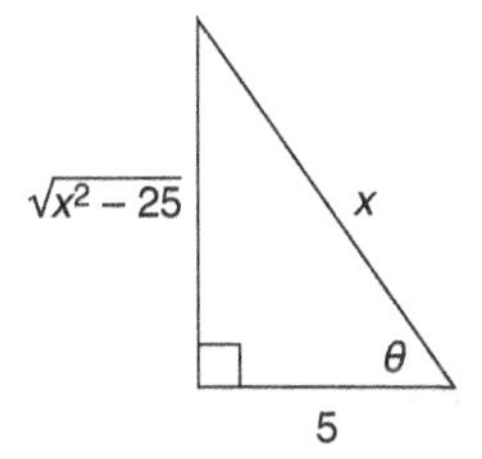

Third side $= \sqrt{x^2 - 5^2} = \sqrt{x^2 - 25}$.

Using the sides of the right triangle, find $\tan(\theta)$:

$$\tan(\theta) = \frac{\text{opposite}}{\text{adjacent}} = \frac{\sqrt{x^2 - 25}}{5}$$

Substitute into the solution:

$$\int \frac{\sqrt{x^2 - 25}}{x} dx = 5\tan(\theta) - 5\theta + C$$

$$= 5 \bullet \frac{\sqrt{x^2 - 25}}{5} - 5 \bullet \sec^{-1}\left(\frac{x}{5}\right) + C$$

$$= \sqrt{x^2 - 25} - 5 \bullet \sec^{-1}\left(\frac{x}{5}\right) + C$$

When asked to evaluate an integral, think of the integration tools much like the order of operations. Start with the first tool that was discussed. If that tool doesn't meet the criteria, then move to the next technique. The order you should consider is as follows:

1. Power Rule for Integration of Polynomials
2. Integration Formulas
3. *U*-Substitution
4. Integration by Parts
5. Partial Fraction Decomposition
6. Complete the Square
7. Powers of Trigonometric Functions
8. Trigonometric Substitutions

BRAIN TICKLERS Set # 28

1. Evaluate $\int \frac{dx}{x^2 + 4x + 13}$.
2. Evaluate $\int \frac{dy}{\sqrt{6y - y^2}}$.
3. Evaluate $\int \tan^2(x)dx$.
4. Evaluate $\int \frac{dx}{\sqrt{25 + x^2}}$.

(Answers are on page 208.)

BRAIN TICKLERS—THE ANSWERS

Set # 23, page 162

1. $\dfrac{t^3}{3} + t^{-1} + C = \dfrac{t^3}{3} + \dfrac{1}{t} + C$
2. $-\cot(x) + C$
3. $\dfrac{1}{2}\ln|x| + C$
4. $3\tan^{-1}(x) + C$

Set # 24, page 170

1. $\dfrac{1}{18}(1-8x^3)^{-3} + C = \dfrac{1}{18(1-8x^3)^3} + C$
2. $\cos(\cos(x)) + C$
3. $\ln|\tan x| + C$
4. $\ln|x^5 + 1| + C$

Set # 25, page 175

1. $\dfrac{1}{2}\ln|x^2 + 1| + C$
2. $10\tan^{-1}(x) + C$
3. $8\tan^{-1}\left(\dfrac{x}{5}\right) + C$
4. $\dfrac{1}{4}\sin^{-1}\left(\dfrac{4x}{3}\right) + C$

Set # 26, page 185

1. $xe^x - e^x + C$
2. $\dfrac{2t\sin(3t)}{3} + \dfrac{2\cos(3t)}{9} + C$
3. $\dfrac{e^{-x}\sin(x)}{2} - \dfrac{e^{-x}\cos(x)}{2} + C$
4. $x\ln(x^3) - 3x + C$

Set # 27, page 194

1. $-\dfrac{1}{4}\ln|x+3| + \dfrac{1}{4}\ln|x-1| + C$
2. $\ln|x| - \ln|2x+1| + C$
3. $-2\ln|x-2| + 3\ln|x-3| + C$
4. $\dfrac{3}{2}\ln|2x-1| - 2\ln|x+1| + C$

Set # 28, page 206

1. $\frac{1}{3}\tan^{-1}\left(\frac{x+2}{3}\right)+C$
2. $\sin^{-1}\left(\frac{y-3}{3}\right)+C$
3. $\tan(x)-x+C$
4. $\ln\left|\frac{\sqrt{25+x^2}}{5}+\frac{x}{5}\right|+C$

(Hint: See Example 15 on page 169)

Chapter 6

Definite Integrals

Just like derivatives, integrals have graphical meanings and applications. The definite integral of a function is different from an indefinite integral. The definite integral has start and end values and is a number that represents the area under the curve of $f(x)$ over an interval.

Notation

A definite integral is like an indefinite integral with respect to the integral symbol and the integrand. Additional limits or bounds are written at the bottom and top of the integral symbol, which represents the interval for the area under the curve.

$$\int_a^b f(x)\,dx = A$$

The above is read as the definite integral of $f(x)$ with respect to x from a to b is equal to A. As shown in the figure below, A represents the value of the area under the curve $f(x)$, bounded by the limits a and b and by the x-axis.

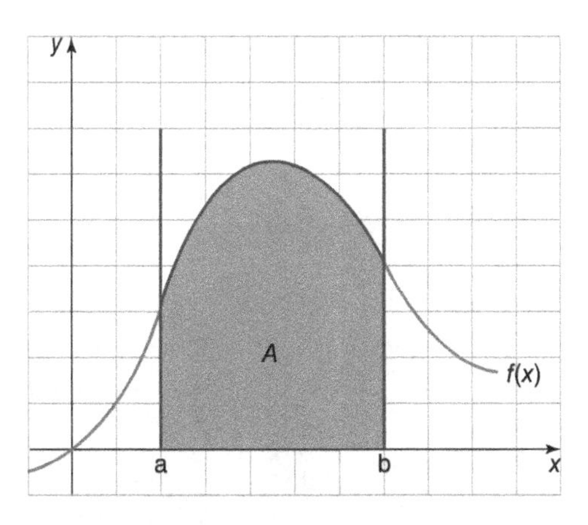

1+2=3 MATH TALK!

The major difference between an indefinite integral and a definite integral is what their answers represent. The indefinite integral of $f(x)$ is a *function* that represents the antiderivative of its integrand. The definite integral of $f(x)$ is a *number value* that represents the area under $f(x)$ bounded by limits and the x-axis.

Example 1:

Use formulas from geometry to evaluate each definite integral:

1. $\int_2^5 x\,dx$

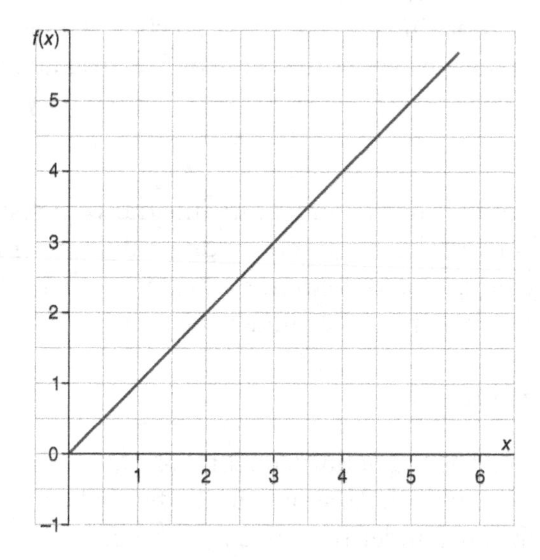

2. $\int_0^3 \sqrt{9 - x^2}\,dx$

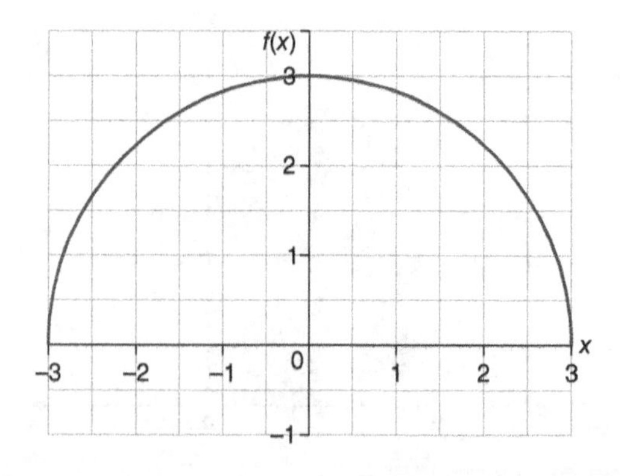

Solution:

1. For the given graph and definite integral, first highlight the region for which the area will be found by drawing in the boundaries.

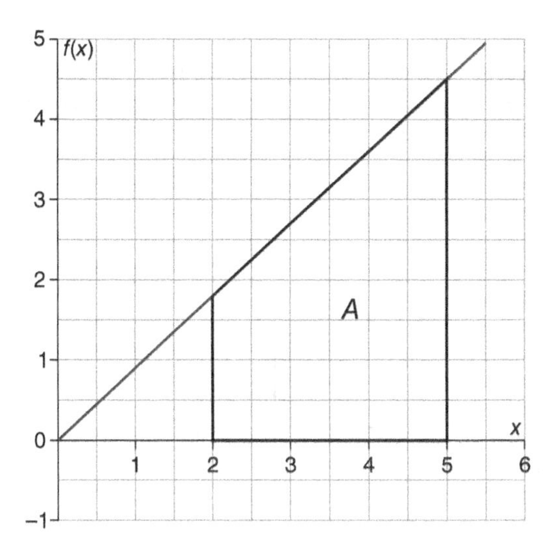

The region that was formed by the function, boundaries, and x-axis is a trapezoid. The definite integral can be evaluated using the area formula for a trapezoid, $\frac{(\text{base } 1 + \text{base } 2) \bullet \text{height}}{2}$.

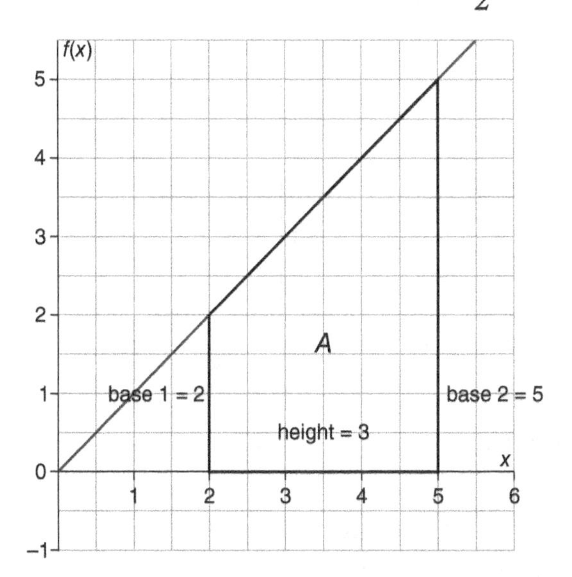

Calculate the area of A:

$$\frac{(2+5)\bullet 3}{2}=\frac{21}{2}$$

$$\int_{2}^{5} x\,dx=\frac{21}{2}$$

2. For the given graph and definite integral, first highlight the region for which the area will be found by drawing in the boundaries.

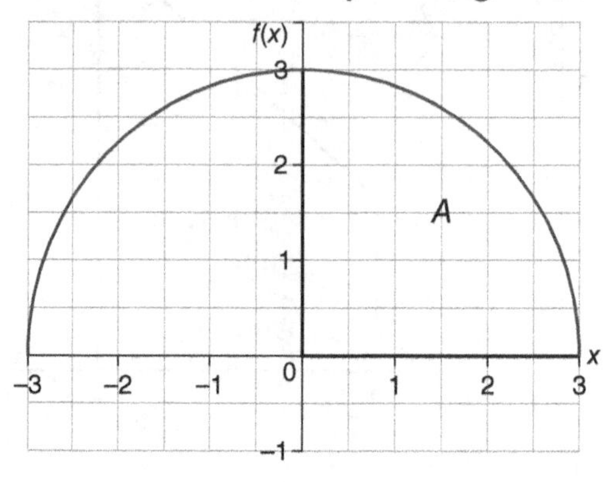

The region that was formed by the function, boundaries, and x-axis is a quarter-circle.

The definite integral can be evaluated using the area formula for a quarter-circle, $\frac{\pi r^2}{4}$:

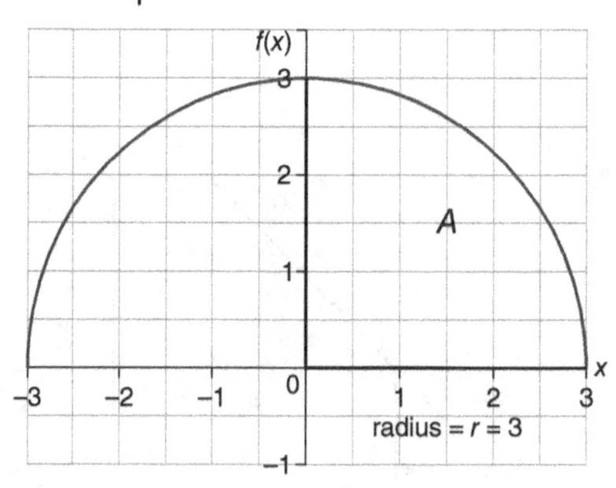

Calculate the area of A:

$$\frac{\pi(3)^2}{4}=\frac{9\pi}{4}$$

$$\int_{0}^{3}\sqrt{9-x^2}\,dx=\frac{9\pi}{4}$$

Properties of Definite Integrals

There are seven definite integral properties that can be applied to evaluating definite integrals. Let a, b, and c represent real numbers.

Property 1: $\int_a^a f(x)\,dx = 0$

Taking the integral at a point results in no area under the curve; therefore, the integral is zero.

Property 2: $\int_a^b f(x)\,dx = -\int_b^a f(x)\,dx$

Switching the bounds of the integral negates the integral.

Property 3: $\int_a^b cf(x)\,dx = c\int_a^b f(x)\,dx$

A constant can be taken out of the integrand and multiplied by the value of the integral.

Property 4: $\int_a^b [f(x) \pm g(x)]\,dx = \int_a^b f(x)\,dx \pm \int_a^b g(x)\,dx$

The sum or difference of two functions in the integrand is equivalent to the sum or difference of the two integrals of the functions.

Property 5: If $f(x) \geq 0$ on $[a, b]$, then $\int_a^b f(x)\,dx \geq 0$

If the function of the integrand is above the x-axis over the interval, the value of the integral is positive.

Property 6: If $f(x) \leq 0$ on $[a, b]$, then $\int_a^b f(x)\,dx \leq 0$

If the function of the integrand is below the x-axis over the interval, the value of the integral is negative.

Property 7: If a, c, and b are any three points on a closed interval, as shown in the diagram, then

$$\int_a^b f(x)\,dx = \int_a^c f(x)\,dx + \int_c^b f(x)\,dx.$$

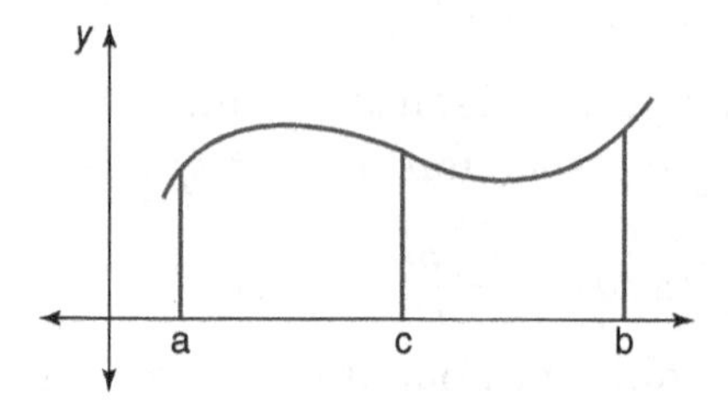

Example 2:

Use formulas from geometry and the graph below to evaluate the definite integral $\int_{-2}^{4} f(x)\,dx$.

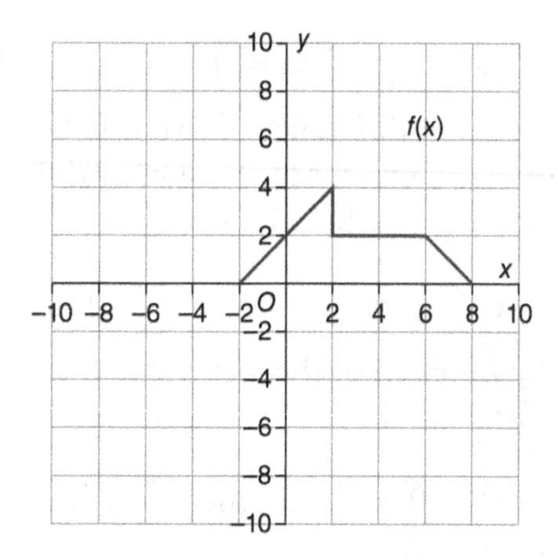

Solution:

For the given graph and definite integral, first highlight the region for which the area will be found by drawing in the boundaries.

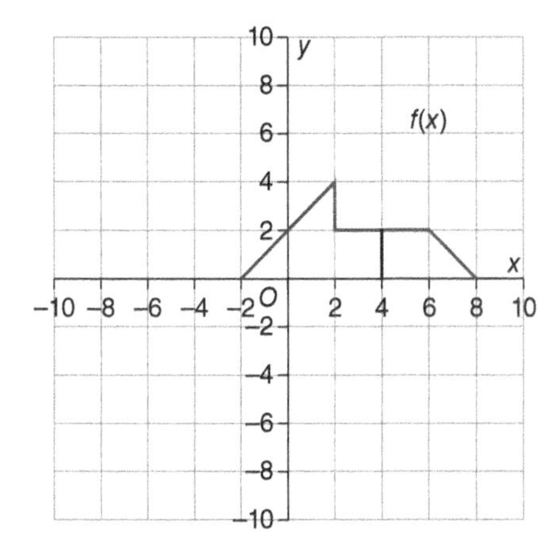

The region that was formed by the function, boundaries, and x-axis has multiple shapes. Using Property 7, the region can be broken into two definite integrals that form geometric shapes:

$$\int_{-2}^{4} f(x)\,dx = \int_{-2}^{2} f(x)\,dx + \int_{2}^{4} f(x)\,dx$$

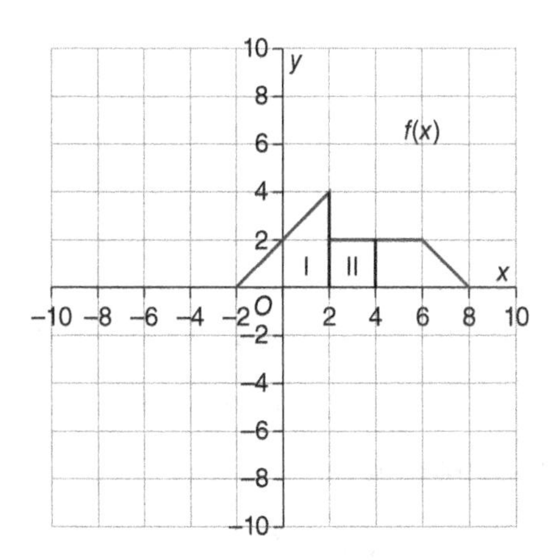

The first definite integral is represented by a triangle and can be evaluated using the formula $\frac{\text{base} \bullet \text{height}}{2}$. The second definite integral is represented by a square and can be evaluated using the formula (side)^2:

$$\text{Area of I} = \frac{4 \cdot 4}{2} = \frac{16}{2} = 8$$

$$\text{Area of II} = (2)^2 = 4$$

$$\int_{-2}^{4} f(x)\,dx = \int_{-2}^{2} f(x)\,dx + \int_{2}^{4} f(x)\,dx = 8 + 4 = 12$$

Example 3:

Evaluate each definite integral using geometric formulas and the diagram below:

1. $\int_{5}^{6} f(x)\,dx$

2. $\int_{6}^{5} f(x)\,dx$

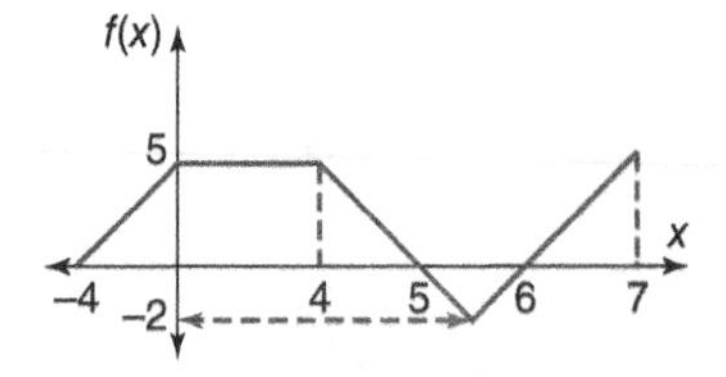

Solution:

1. Since the shape in the interval $[5, 6]$ is a triangle, use the formula $\frac{\text{base} \cdot \text{height}}{2}$. The base equals 1, and the height equals 2. The area can be calculated:

$$\frac{\text{base} \cdot \text{height}}{2} = \frac{1 \cdot 2}{2} = 1$$

By Property 6, since the function is below the x-axis in this interval, the integral is negative:

$$\int_{5}^{6} f(x)\,dx = -1$$

2. Typically, intervals go from a smaller value to a greater value. To rewrite the interval, use Property 2:

$$\int_6^5 f(x)\,dx = -\int_5^6 f(x)\,dx$$

Since this is the same integral as in Part 1:

$$\int_6^5 f(x)\,dx = -\int_5^6 f(x)\,dx = -(-1) = 1$$

BRAIN TICKLERS Set # 29

1. Evaluate: $\int_4^4 f(x)\,dx$.
2. Evaluate using geometric formulas: $\int_1^4 3\,dx$.

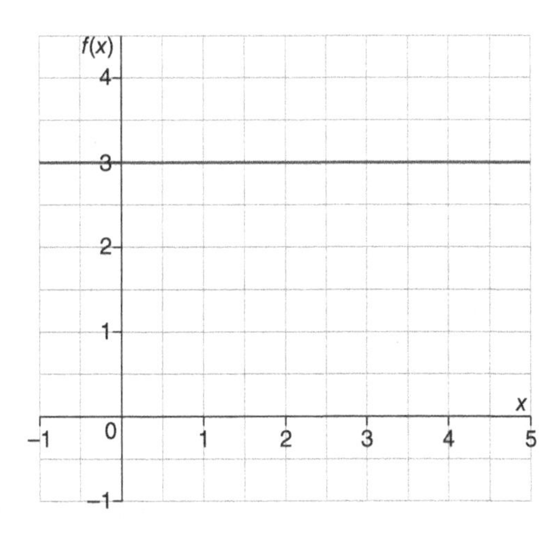

3. Evaluate using geometric formulas: $\int_0^{-4} f(x)\,dx$.

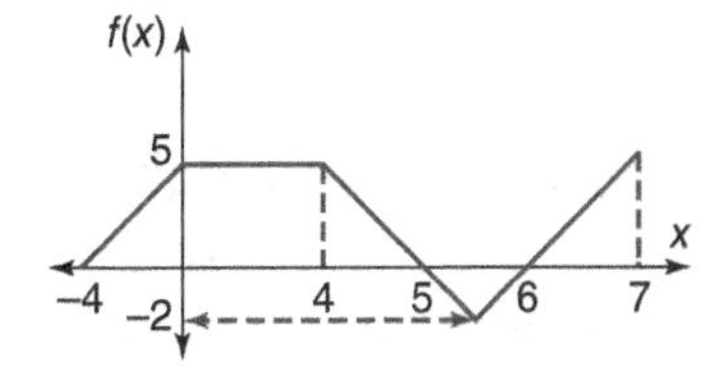

4. Evaluate using geometric formulas: $\int_{-3}^{8} h(x)\,dx$.

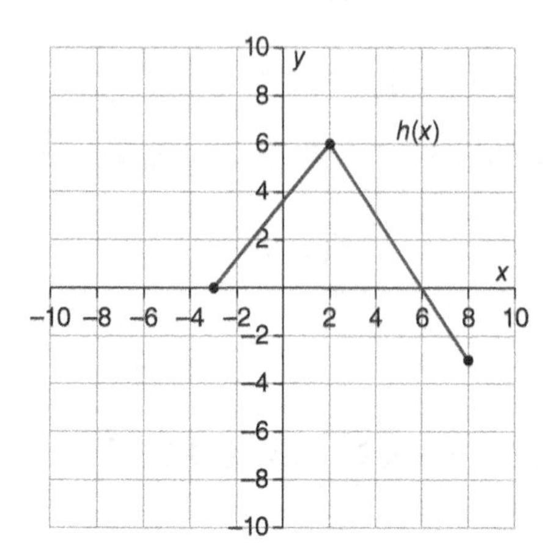

(Answers are on page 250.)

So far, the definite integrals have graphed shapes where the area could be found using geometric formulas. Not all functions, though, form graphs of common shapes. One way to approximate the area under the curve is to use common shapes to fill in the area under the curve, such as rectangles or trapezoids. Then calculate the area of these shapes and find their sum. This method is known as the Riemann sum or Trapezoidal Sum.

Riemann Sum

A Riemann sum is a way to approximate the area under a curve by drawing in rectangles under the curve, calculating the area of each rectangle, and then finding the sum of all the rectangular areas. The more rectangles that are drawn in under the curve, the closer the approximation is to the actual area under the curve.

There are three ways to draw in the rectangles and to find their approximations: a left Riemann sum, a right Riemann sum, and a midpoint Riemann sum. These are shown below for a function $f(x)$ along with the actual area under the curve. Depending upon the curve and which Riemann sum is chosen, the approximation can be an overestimate or an underestimate from the actual value.

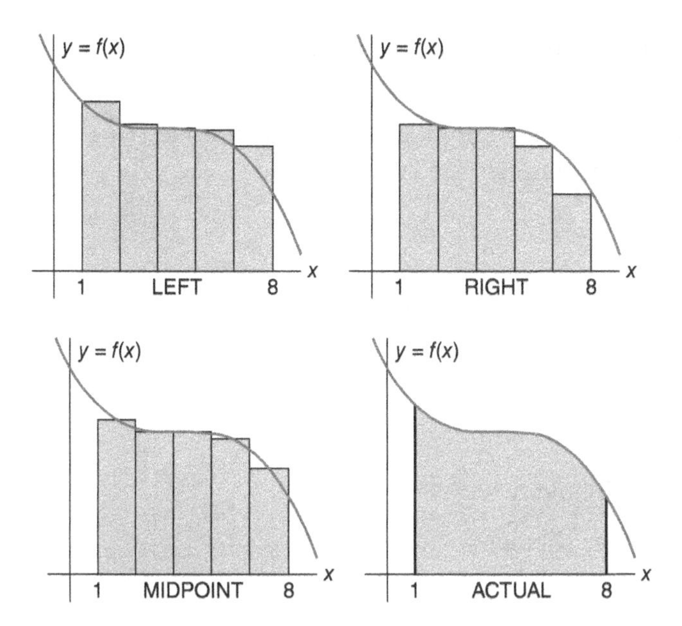

1+2=3 MATH TALK!

The left, right, or midpoint Riemann sum implies how the height of each rectangle is found. For a left Riemann sum, the left endpoint of each interval determines the height of the rectangle. For a right Riemann sum, the right endpoint of each interval determines the height of the rectangle. For a midpoint Riemann sum, the midpoint of the interval determines the height of the rectangle.

Finding the area under the curve by Riemann sum is *painless*. It follows these five steps.

Step 1: Graph the given function.

Step 2: Divide the x-axis into the given number of subintervals. This length is also referred to as Δx. This represents the base of each rectangle.

Step 3: Connect the point on the x-axis (this x-value will be the left, right, or midpoint of each subinterval) with its corresponding y-value on the curve with a straight-line segment. This represents the height of each rectangle. Another way to determine the height of each rectangle is to substitute the appropriate x-value into the function.

Step 4: For each rectangle, calculate the area by finding the product of the base and the height.

Step 5: Find the sum of all the rectangular areas. This approximates the area under the curve.

Example 4:

For each of the following specified below, use a Riemann sum with three subintervals of equal length to approximate the area under the curve for $f(x) = x^2$ from $x = 0$ to $x = 3$.

1. Left Riemann sum
2. Right Riemann sum
3. Midpoint Riemann sum

Solution:

1. Determine the left Riemann sum.

Step 1: Graph the given function.

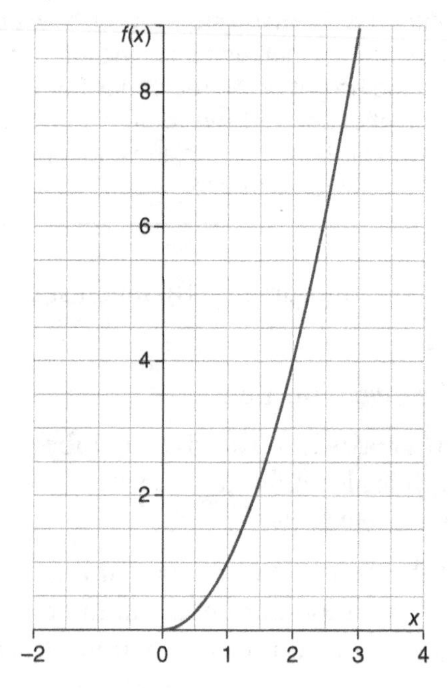

Step 2: Divide the x-axis into the given number of subintervals.

Since the interval is from $x = 0$ to $x = 3$, each subinterval will have a length of 1 along the x-axis, $\Delta x = 1$. The base of each rectangle has a length of 1.

Step 3: Connect the point on the x-axis (this x-value will be the left, right, or midpoint of each subinterval) with its corresponding y-value on the curve with a straight-line segment.

The three equally spaced subintervals are $[0, 1]$, $[1, 2]$, and $[2, 3]$. Since this is a left Riemann sum, the left endpoint of each interval will be the x-value used to construct the segment that connects to the curve.

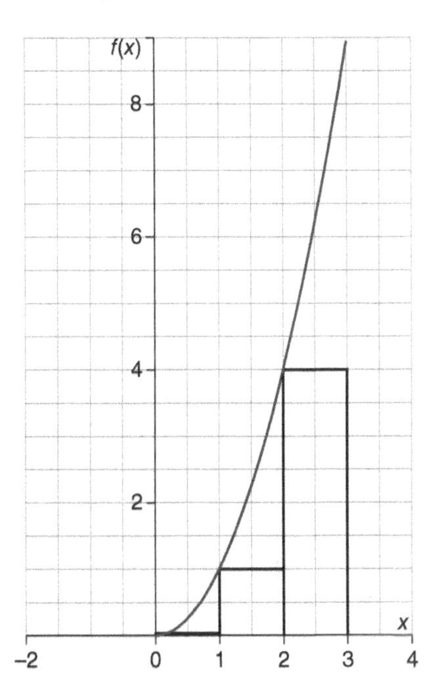

To determine the height of each rectangle, substitute the x-value into the function $f(x) = x^2$:

The height of rectangle 1 $= (0)^2 = 0$

The height of rectangle 2 $= (1)^2 = 1$

The height of rectangle 3 $= (2)^2 = 4$

Step 4: For each rectangle, calculate the area by finding the product of the base and the height:

The area of rectangle 1 $= (1)(0) = 0$

The area of rectangle 2 $= (1)(1) = 1$

The area of rectangle 3 $= (1)(4) = 4$

Step 5: Find the sum of all the rectangular areas. This approximates the area under the curve:

Sum of rectangular areas $= 0 + 1 + 4 = 5$

Area under curve ≈ 5

2. Determine the right Riemann sum.

Step 1: Graph the given function.

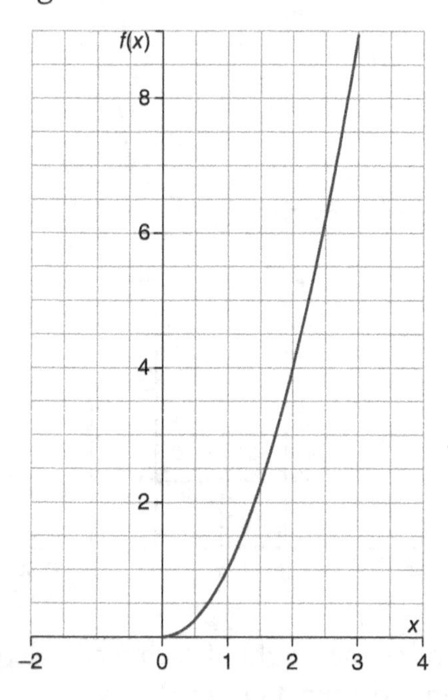

Step 2: Divide the x-axis into the given number of subintervals.

Since the interval is from $x = 0$ to $x = 3$, each subinterval will have a length of 1 along the x-axis, $\Delta x = 1$. The base of each rectangle has a length of 1.

Step 3: Connect the point on the x-axis (this x-value will be the left, right, or midpoint of each subinterval) with its corresponding y-value on the curve with a straight-line segment.

The three equally spaced subintervals are $[0, 1]$, $[1, 2]$, and $[2, 3]$. Since this is a right Riemann sum, the right endpoint of each interval will be the x-value used to construct the segment that connects to the curve.

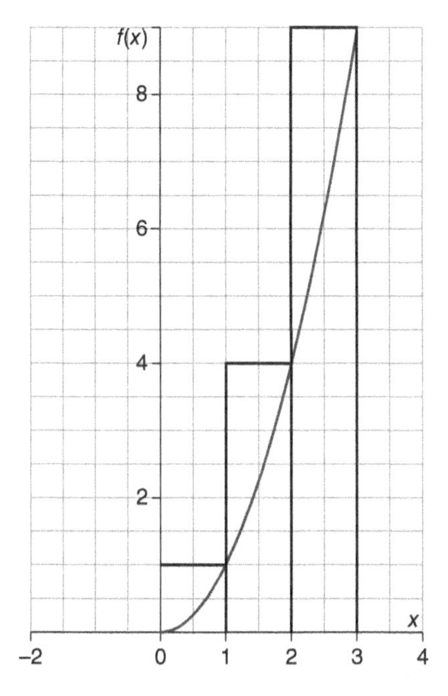

To determine the height of each rectangle, substitute the x-value into the function $f(x) = x^2$:

The height of rectangle 1 $= (1)^2 = 1$

The height of rectangle 2 $= (2)^2 = 4$

The height of rectangle 3 $= (3)^2 = 9$

Step 4: For each rectangle, calculate the area by finding the product of the base and the height:

The area of rectangle 1 $= (1)(1) = 1$

The area of rectangle 2 $= (1)(4) = 4$

The area of rectangle 3 $= (1)(9) = 9$

Step 5: Find the sum of all the rectangular areas. This approximates the area under the curve:

Sum of rectangular areas $= 1 + 4 + 9 = 14$

Area under curve ≈ 14

3. Determine the midpoint Riemann sum.

Step 1: Graph the given function.

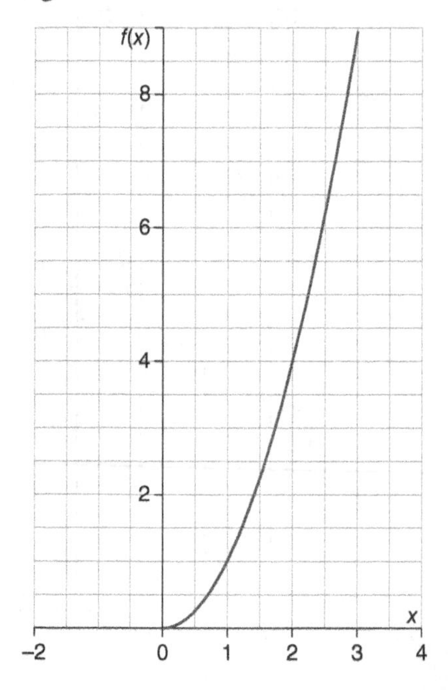

Step 2: Divide the x-axis into the given number of subintervals.

Since the interval is from $x = 0$ to $x = 3$, each subinterval will have a length of 1 along the x-axis, $\Delta x = 1$. The base of each rectangle has a length of 1.

Step 3: Connect the point on the x-axis (this x-value will be the left, right, or midpoint of each subinterval) with its corresponding y-value on the curve with a straight-line segment.

The three equally spaced subintervals are $[0, 1]$, $[1, 2]$, and $[2, 3]$. Since this is a midpoint Riemann sum, the midpoint of each interval will be the x-value used to construct the segment that connects to the curve.

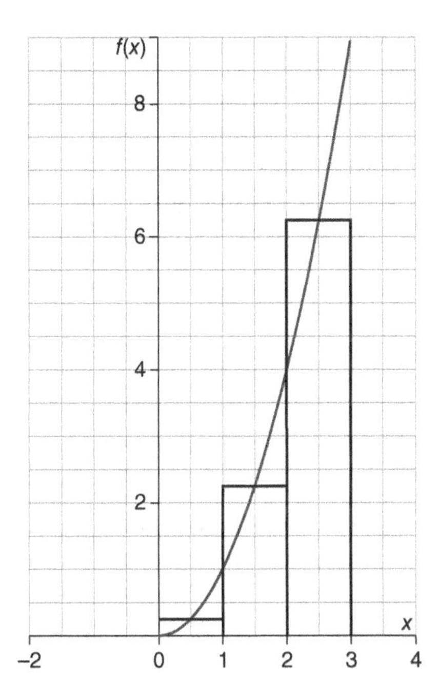

To determine the height of each rectangle, substitute the x-value into the function $f(x) = x^2$:

The height of rectangle 1 $= (0.5)^2 = 0.25$

The height of rectangle 2 $= (1.5)^2 = 2.25$

The height of rectangle 3 $= (2.5)^2 = 6.25$

Step 4: For each rectangle, calculate the area by finding the product of the base and the height:

The area of rectangle 1 $= (1)(0.25) = 0.25$

The area of rectangle 2 $= (1)(2.25) = 2.25$

The area of rectangle 3 $= (1)(6.25) = 6.25$

Step 5: Find the sum of all the rectangular areas. This approximates the area under the curve:

Sum of rectangular areas $= 0.25 + 2.25 + 6.25 = 8.75$

Area under curve ≈ 8.75

Look at the three different approximations in this case; the left Riemann sum was an underestimate, and the right Riemann sum was an overestimate. All three are considered approximations and are not the actual area under the curve. If the number of subintervals increases, the widths of the rectangles will get smaller. The approximation of the area under the curve will get better as the thinner rectangles fill in the gaps under the curve and the portions that overlap the curve diminish. Having infinitely many subintervals allows for the following equation:

$$\int_a^b f(x)\,dx = \lim_{n\to\infty}\sum_{i=1}^{n} f(x_i)\Delta x$$

This means that the area under the curve, which is the definite integral, is equal to the infinite sum of the areas of the rectangles where the height is $f(x_i)$ and the length is Δx.

Example 5:

Let $f(x)$ be a function that is twice differentiable for all real numbers. The table below gives values of $f(x)$ for selected points in the closed interval $[3, 14]$. Use a left Riemann sum with subintervals indicated by the data in the table to approximate $\int_3^{14} f(x)\,dx$.

x	3	5	8	10	14
f(x)	4	−2	3	6	11

Solution:
Since the function equation was not given, move to Step 2 to find the left Riemann sum.

Step 2: Divide the x-axis into the given number of subintervals. This length is also referred to as Δx. This represents the base of each rectangle.

The subintervals indicated by the table are $[3, 5]$, $[5, 8]$, $[8, 10]$, and $[10, 14]$:

Base of rectangle 1 $= 5 - 3 = 2$

Base of rectangle 2 $= 8 - 5 = 3$

Base of rectangle 3 $= 10 - 8 = 2$

Base of rectangle 4 $= 14 - 10 = 4$

Step 3: Connect the point on the x-axis (this x-value will be the left, right, or midpoint of each subinterval) with its corresponding y-value on the curve with a straight-line segment. This represents the height of each rectangle.

Another way to determine the height of each rectangle is to use the table of values to determine the y-value for each corresponding x-value. Since this is a left Riemann sum, the left endpoint of each interval will determine the height:

Height of rectangle $1 = 4$

Height of rectangle $2 = -2$

Height of rectangle $3 = 3$

Height of rectangle $4 = 6$

Step 4: For each rectangle, calculate the area by finding the product of the base and the height:

Area of rectangle $1 = (2)(4) = 8$

Area of rectangle $2 = (3)(-2) = -6$

Area of rectangle $3 = (2)(3) = 6$

Area of rectangle $4 = (4)(6) = 24$

Step 5: Find the sum of all the rectangular areas. This approximates the area under the curve:

Sum of rectangular areas $= 8 + -6 + 6 + 24 = 32$

Area under curve ≈ 32

Since the Riemann sum is an alternative way of finding the definite integral, it can also be used when an antiderivative is needed.

Example 6:

A cylindrical water tower is being filled with water. During the time interval $0 \leq t \leq 12$ hours, water is pumped into the tower at the rate of $W(t)$ cubic feet per hour. The table below gives values of $W(t)$. Use a midpoint Riemann sum with three subintervals of equal length to approximate the total amount of water that is pumped into the tower during the time interval $0 \leq t \leq 12$ hours.

t	0	2	4	6	8	10	12
$W(t)$	0	23	46	53	58	61	65

Solution:

The given function represents the rate at which the water is pumped into the tower. To find the amount of water in the tower, the antiderivative, or definite integral of the function over the interval, must be found. To evaluate the definite integral, the problem states to use a midpoint Riemann approximation.

Since the function equation was not given, move to Step 2 to find the midpoint Riemann sum.

Step 2: Divide the x-axis into the given number of subintervals. This length is also referred to as Δx. This represents the base of each rectangle.

The subintervals indicated by the table are $[0, 4]$, $[4, 8]$, and $[8, 12]$:

Base of rectangle $1 = 4 - 0 = 4$

Base of rectangle $2 = 8 - 4 = 4$

Base of rectangle $3 = 12 - 8 = 4$

Step 3: Connect the point on the x-axis (this x-value will be the left, right, or midpoint of each subinterval) with its corresponding y-value on the curve with a straight-line segment. This represents the height of each rectangle.

Another way to determine the height of each rectangle is to use the table of values to determine the y-value for each corresponding x-value. Since this is a midpoint Riemann sum, the midpoint of each interval will determine the height:

Height of rectangle $1 = 23$

Height of rectangle $2 = 53$

Height of rectangle $3 = 61$

Step 4: For each rectangle, calculate the area by finding the product of the base and the height:

Area of rectangle 1 $= (4)(23) = 92$

Area of rectangle 2 $= (4)(53) = 212$

Area of rectangle 3 $= (4)(61) = 244$

Step 5: Find the sum of all the rectangular areas. This approximates the area under the curve:

Sum of rectangular areas $= 92 + 212 + 244 = 548$

Area under curve ≈ 548

The total amount of water that is pumped into the tower is approximately 548 cubic feet.

Trapezoidal Sum

The Riemann sum is a useful approximation because using the formula for the area of a rectangle is simple. However, the shape of a rectangle is not necessarily the best fit for a curve. Another approximation tool is the trapezoidal sum. Instead of fitting rectangles under the curve, the trapezoid is used as its legs can slant to fit the curve. The more trapezoids that are used, the better the fit and the closer the approximation is to the actual area under the curve.

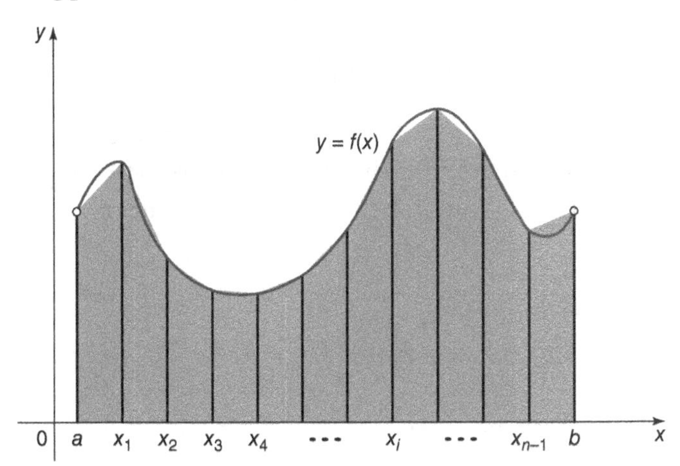

The formula to find the area of a trapezoid is the following:

$$\text{area} = \frac{(\text{base}_1 + \text{base}_2) \bullet \text{height}}{2} = \frac{(b_1 + b_2) \bullet h}{2} = \frac{h}{2} \bullet (b_1 + b_2)$$

The bases of each trapezoid are determined by the function value, or y-value, of each endpoint in the interval. For example, in the diagram above for the first trapezoid, $\text{base}_1 = f(a)$ and $\text{base}_2 = f(x_1)$.

The height of each trapezoid is determined by the difference of each endpoint in the interval, which is also Δx. For example, in the diagram above for the first trapezoid, $\Delta x = x_1 - a$.

In general, to find the trapezoidal sum if the continuous function is divided into n even subintervals:

$$\Delta x = \frac{b - a}{n} = h$$

The sum of the n trapezoids can be calculated:

$$\text{Sum} = \frac{h}{2}[\text{base}_1 + \text{base}_2 + \text{base}_2 + \cdots + \text{base}_{n-1} + \text{base}_{n-1} + \text{base}_n]$$

$$= \frac{\Delta x}{2} \bullet [f(a) + f(x_1) + f(x_1) + \cdots + f(x_{n-1}) + f(x_{n-1}) + f(b)]$$

$$= \frac{b - a}{2n} \bullet [f(a) + 2f(x_1) + 2f(x_2) + \cdots + 2f(x_{n-1}) + f(b)]$$

This sum approximates the area under the curve.

Taking the limit as n approaches infinity means there are infinitely many trapezoids fitting under the curve. This would be equal to the definite integral and allows for the following equation:

$$\int_a^b f(x)\,dx = \lim_{n \to \infty}\left[\frac{b - a}{2n} \bullet [f(a) + 2f(x_1) + 2f(x_2) + \ldots + 2f(x_{n-1}) + f(b)]\right]$$

1+2=3 MATH TALK!

The Trapezoidal Rule is like a Riemann sum because it uses geometric formulas of shapes to approximate the area under a curve. Like a Riemann sum, to follow a Trapezoidal Rule you can find the area of each trapezoid in the interval and find the sum to approximate the area. Alternatively, you can use the formula given above. The endpoint function values are not multiplied twice because those are bases that do not connect two trapezoids together.

Finding the area under the curve by the Trapezoidal Rule is *painless.* It follows these five steps.

Step 1: Graph the given function.

Step 2: Divide the x-axis into the given number of subintervals. This length is also referred to as Δx. This represents the height of each trapezoid.

Step 3: Connect the endpoints of each interval from the x-axis to its corresponding y-value on the curve with a straight-line segment. This represents the bases of each trapezoid. Another way to determine the bases of the trapezoids is to substitute the x-value into the function.

Step 4: Use the area formula for each trapezoid.

Step 5: Find the sum of all the trapezoidal areas. This approximates the area under the curve.

Example 7:

Using the Trapezoidal Rule, find the area under the curve $f(x) = x^3$ bounded by the x-axis and in the interval $[1, 3]$ using four trapezoids.

Solution:

Step 1: Graph the given function.

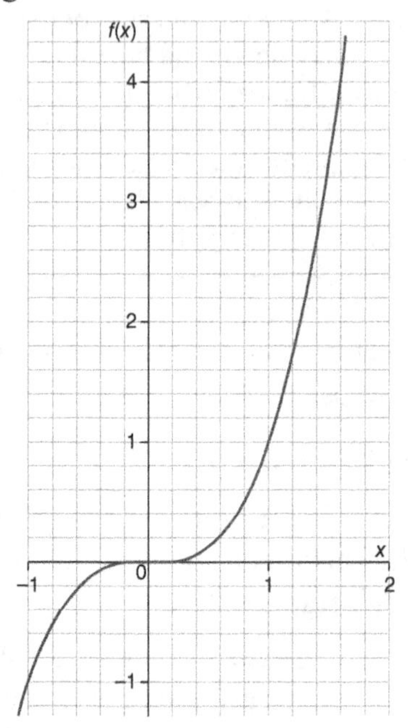

Step 2: Divide the x-axis into the given number of subintervals. This length is also referred to as Δx. This represents the height of each trapezoid.

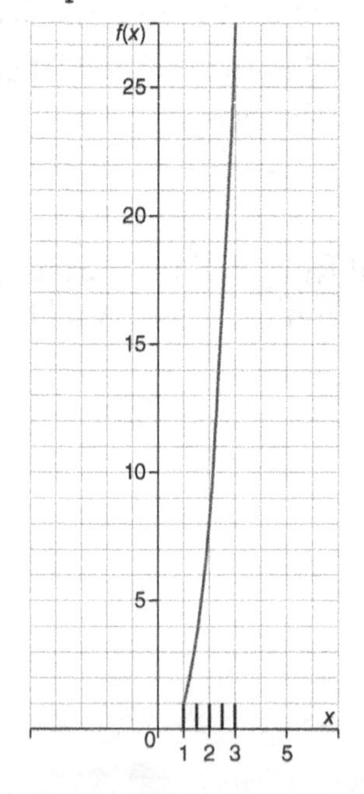

The interval is $[1, 3]$, which means $a = 1$ and $b = 3$. There will be four trapezoids, which means $n = 4$. Calculate the height:

$$\Delta x = \frac{b - a}{n} = \frac{3 - 1}{4} = \frac{1}{2}$$

The height of each trapezoid is $\frac{1}{2} = 0.5$.

Step 3: Connect the endpoints of each interval from the x-axis to its corresponding y-value on the curve with a straight-line segment. This represents the bases of each trapezoid. Another way to determine the bases of the trapezoids is to substitute the x-value into the function.

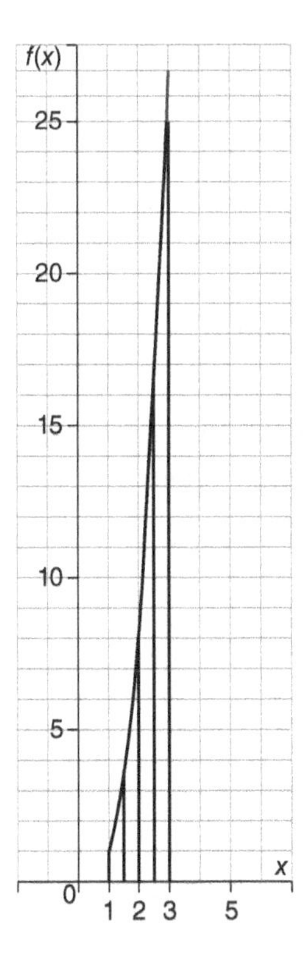

The intervals for the four trapezoids are [1, 1.5], [1.5, 2], [2, 2.5], and [2.5, 3]. Find the bases:

$$f(1) = (1)^3 = 1$$

$$f(1.5) = (1.5)^3 = 3.375$$

$$f(2) = (2)^3 = 8$$

$$f(2.5) = (2.5)^3 = 15.625$$

$$f(3) = (3)^3 = 27$$

Step 4: Use the area formula for each trapezoid:

$$\text{Area of trapezoid 1} = \frac{(1 + 3.375) \bullet 0.5}{2} = 1.09375$$

$$\text{Area of trapezoid 2} = \frac{(3.375 + 8) \bullet 0.5}{2} = 2.84375$$

$$\text{Area of trapezoid 3} = \frac{(8 + 15.625) \bullet 0.5}{2} = 5.90625$$

$$\text{Area of trapezoid 4} = \frac{(15.625 + 27) \bullet 0.5}{2} = 10.65625$$

Step 5: Find the sum of all the trapezoidal areas. This approximates the area under the curve:

$$\text{Sum of trapezoidal areas} = 1.09375 + 2.84375 + 5.90625 + 10.65625 = 20.5$$

Alternatively, you can use the Trapezoidal Rule formula:

$$\text{Area} = \frac{\frac{1}{2}}{2}[f(1) + 2f(1.5) + 2f(2) + 2f(2.5) + f(3)]$$

$$= \frac{1}{4}[1 + 2 \bullet 3.375 + 2 \bullet 8 + 2 \bullet 15.625 + 27] = \frac{1}{4}[82] = 20.5$$

Example 8:

Using the values in the table below, use a trapezoidal sum with five equal subintervals indicated by the table to approximate $\int_0^{15} C(t)\,dt$.

t	0	3	6	9	12	15
C(*t*)	20	31	28	24	22	21

Solution:

Since the function equation was not given, move to Step 2 to find the Trapezoidal Rule.

Step 2: Divide the x-axis into the given number of subintervals. This length is also referred to as Δx. This represents the height of each trapezoid:

$$h = \Delta x = \frac{15 - 0}{5} = 3$$

Step 3: Determine the bases of each trapezoid by using the table of values to determine the y-value for each corresponding x-value.

The intervals for the four trapezoids are $[0, 3]$, $[3, 6]$, $[6, 9]$, $[9, 12]$, and $[12, 15]$. Calculate the bases:

$$C(0) = 20$$
$$C(3) = 31$$
$$C(6) = 28$$
$$C(9) = 24$$
$$C(12) = 22$$
$$C(15) = 21$$

Step 4: Use the area formula for each trapezoid:

$$\text{Area of trapezoid 1} = \frac{(20 + 31) \cdot 3}{2} = 76.5$$

$$\text{Area of trapezoid 2} = \frac{(31 + 28) \cdot 3}{2} = 88.5$$

$$\text{Area of trapezoid 3} = \frac{(28 + 24) \cdot 3}{2} = 78$$

$$\text{Area of trapezoid 4} = \frac{(24 + 22) \bullet 3}{2} = 69$$

$$\text{Area of trapezoid 5} = \frac{(22 + 21) \bullet 3}{2} = 64.5$$

Step 5: Find the sum of all the trapezoidal areas. This approximates the area under the curve.

$$\begin{aligned}\text{Sum of trapezoidal areas} &= 76.5 + 88.5 + 78 + 69 + 64.5 \\ &= 376.5\end{aligned}$$

Alternatively, you can use the Trapezoidal Rule formula:

$$\text{Area} = \frac{3}{2}[C(0) + 2C(3) + 2C(6) + 2C(9) + 2C(12) + C(15)]$$

$$= \frac{3}{2}[20 + 2 \bullet 31 + 2 \bullet 28 + 2 \bullet 24 + 2 \bullet 22 + 21] = \frac{3}{2}[251] = 376.5$$

BRAIN TICKLERS Set # 30

1. Use a right Riemann sum to estimate the area under the curve from $x = -4$ to $x = 4$ where $\Delta x = 1$ for the curve $f(x) = x^2$.
2. Use a midpoint Riemann sum with three subintervals of equal length to approximate the area under the curve for $f(x) = 9 - x^2$ in the interval [0, 3].
3. Use a left Riemann sum with subintervals indicated by the data in the table below to approximate $\int_3^{14} g(x)\,dx$.

x	3	5	8	10	14
$g(x)$	0	1	4	–3	7

4. Find the trapezoidal approximation of $\int_{-4}^{2} e^{-x}\,dx$ if three equal subdivisions of the interval [–4, 2] are used.

(Answers are on page 250.)

Fundamental Theorem of Calculus

There are two parts to the *Fundamental Theorem of Calculus*, both involving definite integrals.

Fundamental Theorem of Calculus Part I

Let the function f be continuous on the closed interval $[a, b]$ and let g be a function such that $g'(x) = f(x)$ for all x in $[a, b]$:

$$\int_a^b f(t)\,dt = g(b) - g(a) = g(x)\Big|_a^b$$

1+2=3 MATH TALK!

This part of the theorem helps to evaluate definite integrals algebraically. To evaluate a definite integral, find the integral of the integrand and then substitute in the upper and lower bounds. The solution is the difference of the two substitutions.

Evaluating definite integrals algebraically is *painless*. It involves the following three steps.

Step 1: Integrate the integrand.

Step 2: Substitute the upper and lower bounds.

Step 3: Evaluate the difference of the substitutions.

Example 9:

Evaluate each of the definite integrals algebraically.

1. $\int_1^2 x^4\,dx$
2. $\int_{-1}^1 (x^{4/3} + 4x^{1/3})\,dx$
3. $\int_0^2 2x^2\sqrt{x^3+1}\,dx$

Solution:

1. Evaluate the definite integral of $\int_1^2 x^4\,dx$.

Step 1: Integrate the integrand:

$$\int_1^2 x^4\,dx = \frac{x^5}{5}\Bigg|_1^2$$

Step 2: Substitute the upper and lower bounds. This is indicated in Step 1 with the straight line:

$$\frac{(2)^5}{5} - \frac{(1)^5}{5}$$

Step 3: Evaluate the difference of the substitutions.

$$\frac{32}{5} - \frac{1}{5} = \frac{31}{5}$$

2. Evaluate the definite integral of $\int_{-1}^1 (x^{4/3} + 4x^{1/3})\,dx$.

Step 1: Integrate the integrand:

$$\int_{-1}^1 (x^{4/3} + 4x^{1/3})\,dx = \left(\frac{x^{7/3}}{\frac{7}{3}} + 4 \bullet \frac{x^{4/3}}{\frac{4}{3}}\right)\Bigg|_{-1}^1 = \left(\frac{3x^{7/3}}{7} + 3x^{4/3}\right)\Bigg|_{-1}^1$$

Step 2: Substitute the upper and lower bounds:

$$\frac{3(1)^{7/3}}{7} + 3(1)^{4/3} - \left(\frac{3(-1)^{7/3}}{7} + 3(-1)^{4/3}\right)$$

Step 3: Evaluate the difference of the substitutions:

$$\frac{3}{7} + 3 - \left(\frac{-3}{7} + 3\right) = \frac{6}{7} + 0 = \frac{6}{7}$$

3. Evaluate the definite integral of $\int_0^2 2x^2\sqrt{x^3+1}\,dx$.

Step 1: Integrate the integrand:

$$\text{Let } u = x^3 + 1,\ du = 3x^2dx \rightarrow x^2dx = \frac{1}{3}du$$

$$\int_0^2 2x^2\sqrt{x^3+1}\,dx \rightarrow 2\int_{x=0}^{x=2}\sqrt{u}\bullet\frac{1}{3}du = \frac{2}{3}\int_{x=0}^{x=2}\sqrt{u}\,du$$

$$= \frac{2}{3}\frac{u^{3/2}}{\frac{3}{2}}\Bigg|_{x=0}^{x=2} = \frac{4}{9}(x^3+1)^{3/2}\Bigg|_0^2$$

Step 2: Substitute the upper and lower bounds:

$$\frac{4}{9}((2)^3+1)^{3/2} - \frac{4}{9}((0)^3+1)^{3/2}$$

Step 3: Evaluate the difference of the substitutions:

$$\frac{4}{9}(8+1)^{3/2} - \frac{4}{9}(0+1)^{3/2} = \frac{4}{9}(9)^{3/2} - \frac{4}{9}(1)^{3/2}$$

$$= \frac{4}{9}(27) - \frac{4}{9}(1) = 12 - \frac{4}{9} = \frac{104}{9}$$

1+2=3 MATH TALK!

In question 3 of the previous example, you may have noticed that after the *u*-substitution, the bounds were written with an *x*. That is because the intergral was rewritten in terms of *u*. To leave the bounds as a number, they too would have to be rewritten in terms of *u*. This can be accomplished by substituting each *x*-value into the *u*-substitution. For example, in the previous problem for $x = 2$, $u = (2)^3 + 1 = 9$, and for $x = 0$, $u = (0)^3 + 1 = 1$. Then the integral could be written as follows: $\int_0^2 2x^2\sqrt{x^3+1}\,dx \rightarrow 2\int_1^9 \sqrt{u}\bullet\frac{1}{3}du$. This way everything would be written in terms of *u* and the intergal could have been evaluted in terms of *u*. Then there would have been no need to substitute back in for *x*:

$$2\int_1^9 \sqrt{u}\bullet\frac{1}{3}du = \frac{2}{3}\int_1^9 \sqrt{u}\,du = \frac{2}{3}\frac{u^{3/2}}{\frac{3}{2}}\Bigg|_1^9 = \frac{4}{9}u^{3/2}\Bigg|_1^9 = \frac{4}{9}((9)^{3/2} - (1)^{3/2})$$

$$= \frac{4}{9}(27-1) = \frac{4}{9}(26) = \frac{104}{9}$$

The same solution is found. So, either method is appropriate, and the choice is left to the discretion of the reader.

Example 10:

Evaluate $\int_0^{\pi/2} \sin^3 x \cos x \, dx$.

Solution:

Step 1: Integrate the integrand:

Let $u = \sin x$, $du = \cos x \, dx$

$$\int_0^{\pi/2} \sin^3 x \cos x \, dx \rightarrow \int_{x=0}^{x=\pi/2} u^3 \, du = \frac{u^4}{4}\Bigg|_{x=0}^{x=\pi/2} = \frac{(\sin x)^4}{4}\Bigg|_0^{\pi/2}$$

Step 2: Substitute the upper and lower bounds:

$$\frac{\left(\sin\left(\frac{\pi}{2}\right)\right)^4}{4} - \frac{(\sin(0))^4}{4}$$

Step 3: Evaluate the difference of the substitutions:

$$\frac{(1)^4}{4} - \frac{(0)^4}{4} = \frac{1}{4} - 0 = \frac{1}{4}$$

Example 11:

Find the area of the shaded region, shown below, for the function $f(x) = x^2$ from $x = 1$ to $x = 2$.

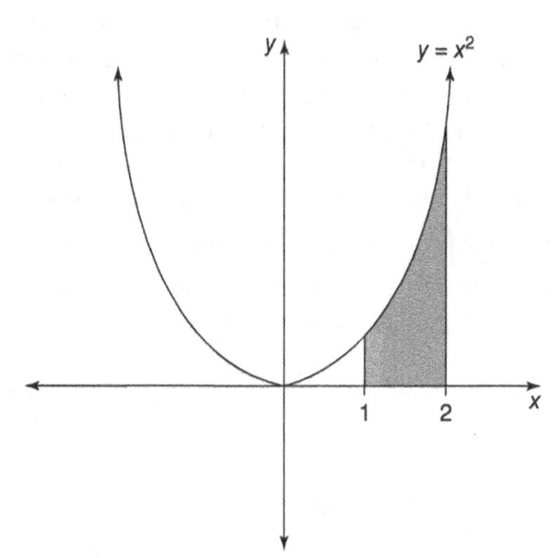

Solution:

The area of the shaded region can be found by evaluating the following definite integral:

$$\int_1^2 x^2\,dx$$

Step 1: Integrate the integrand:

$$\int_1^2 x^2\,dx = \left.\frac{x^3}{3}\right|_1^2$$

Step 2: Substitute the upper and lower bounds:

$$\frac{(2)^3}{3} - \frac{(1)^3}{3}$$

Step 3: Evaluate the difference of the substitutions:

$$\frac{8}{3} - \frac{1}{3} = \frac{7}{3}$$

Fundamental Theorem of Calculus Part II

Let the function f be continuous on the closed interval $[a, b]$ and let x be any number in $[a, b]$:

$$\text{If } A(x) = \int_a^x f(t)\,dt\text{, then } A'(x) = f(x)$$

In other words:

$$\frac{d}{dx}\int_a^x f(t)\,dt = f(x)$$

1+2=3 MATH TALK!

This part of the theorem establishes the relationship between the integral and the derivative. It says that the rate of change (the derivative) of the area under the curve up to a point x (the definite integral) equals the height of the area at that point (the function).

CAUTION—Major Mistake Territory!

To apply this theorem, it is important to note the boundaries of the integral. The lower bound is a constant and can be any real number. The upper bound is the variable *x* with which the derivative is being found with respect to. If the given order of the boundaries is different, the integral must be manipulated to apply the Fundamental Theorem of Calculus.

Example 12:

Compute the following derivatives:

1. $\frac{d}{dx}\int_0^x \sqrt{4+t^2}\,dt$
2. $\frac{d}{dx}\int_1^x \frac{1}{t^2+1}\,dt$
3. $\frac{d}{dx}\int_x^3 \sqrt{1+t^4}\,dt$

Solution:

1. To find the derivative of an integral of this form, use Part II of the Fundamental Theorem of Calculus:

$$\frac{d}{dx}\int_0^x \sqrt{4+t^2}\,dt = \sqrt{4+x^2}$$

2. To find the derivative of an integral of this form, use Part II of the Fundamental Theorem of Calculus:

$$\frac{d}{dx}\int_1^x \frac{1}{t^2+1}\,dt = \frac{1}{x^2+1}$$

3. To be able to apply Part II of the Fundamental Theorem of Calculus, the bounds of the integral must be switched. Using Property 2:

$$\frac{d}{dx}\int_x^3 \sqrt{1+t^4}\,dt = \frac{d}{dx}\left(-\int_3^x \sqrt{1+t^4}\,dt\right) = -\sqrt{1+x^4}$$

Combining the two Fundamental Theorems of Calculus along with the Chain Rule allows us to evaluate derivatives of integrals whose bounds are functions of x.

If u and v are functions of x and if $F(x) = \int_v^u f(t)\,dt$:

$$F'(x) = \frac{d}{dx}\int_v^u f(t)\,dt = f(u) \bullet \frac{du}{dx} - f(v) \bullet \frac{dv}{dx}$$

Example 13:

Find $G'(x)$ if $G(x) = \int_{2x}^{x^3} 5t\,dt$.

Solution:

$$G'(x) = \frac{d}{dx}\int_{2x}^{x^3} 5t\,dt = (5(x^3)) \bullet (3x^2) - (5(2x)) \bullet (2) = 15x^5 - 20x$$

Example 14:

Evaluate $\frac{d}{dx}\int_{-1}^{x^2} (3t + 4)\,dt$.

Solution:

$$\begin{aligned}\frac{d}{dx}\int_{-1}^{x^2} (3t + 4)\,dt &= (3(x^2) + 4) \bullet (2x) - (3(-1) + 4) \bullet (0) \\ &= (3(x^2) + 4) \bullet (2x) = (3x^2 + 4) \bullet (2x) = 6x^3 + 8x\end{aligned}$$

Example 15:

Evaluate $\frac{d}{dx}\int_{x}^{x^2+3x-1} (2t + 7)\,dt$.

Solution:

$$\begin{aligned}\frac{d}{dx}\int_{x}^{x^2+3x-1} (2t + 7)\,dt &= (2(x^2 + 3x - 1) + 7) \bullet (2x + 3) - (2(x) + 7) \bullet (1) \\ &= (2x^2 + 6x - 2 + 7) \bullet (2x + 3) - (2x + 7) \bullet (1) \\ &= (2x^2 + 6x + 5) \bullet (2x + 3) - (2x + 7) \\ &= 4x^3 + 12x^2 + 10x + 6x^2 + 18x + 15 - 2x - 7 \\ &= 4x^3 + 18x^2 + 26x + 8\end{aligned}$$

BRAIN TICKLERS Set # 31

1. Evaluate $\int_0^6 (4x^3 - 2x + 5)\,dx$.
2. Evaluate $\int_{-1}^2 \frac{x^2}{e^{x^3}}\,dx$.
3. Find the area under the curve for the function $f(x) = e^{2x}$, shown below, from $x = 0$ to $x = 1$.

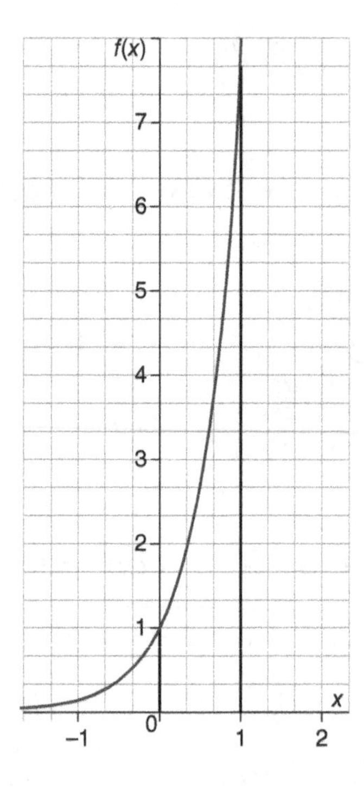

4. If $f(x) = \int_0^{x^4} \frac{1}{\sqrt{t^2 + 5}}\,dt$, find $f'(x)$.

(Answers on page 250.)

Mean Value Theorem for Integrals

If $f(x)$ is continuous on a closed interval $[a, b]$, then there is at least one point c in $[a, b]$ such that $f(c) \cdot (b - a) = \int_a^b f(x)\,dx$.

1+2=3 MATH TALK!

The *Mean Value Theorem for Integrals* states that if a function is continuous, there exists an x-value in the interval (a, b) that corresponds to a function value. This function value represents the height of a rectangle whose area is equal to the area under the curve, the definite integral. The height of the rectangle is $f(c)$, and the base of the rectangle is $b - a$.

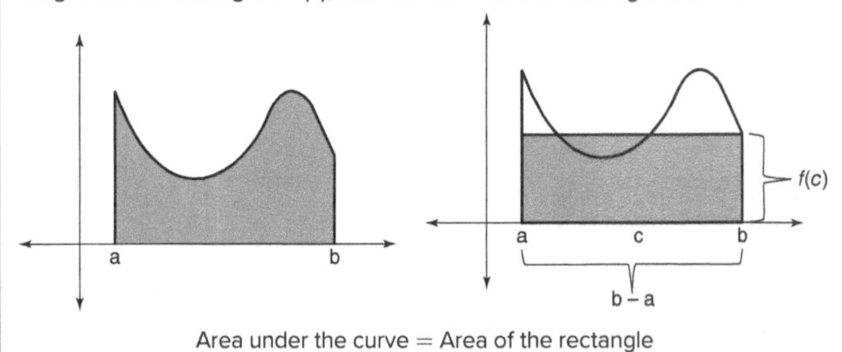

Area under the curve = Area of the rectangle

Example 16:

If $f(x) = x^2$, $a = 1$, $b = 3$, find c guaranteed by the Mean Value Theorem for Integrals.

Solution:

Since $f(x) = x^2$ is a continuous function, by the Mean Value Theorem for Integrals:

$$f(c) \bullet (b - a) = \int_a^b f(x)\,dx$$

$$(c)^2 \bullet (3 - 1) = \int_1^3 x^2\,dx$$

$$2c^2 = \left.\frac{x^3}{3}\right|_1^3$$

$$2c^2 = \frac{(3)^3}{3} - \frac{(1)^3}{3} = \frac{26}{3}$$

$$c^2 = \frac{13}{3}$$

$$c = +\sqrt{\frac{13}{3}}, \text{ where this } c \text{ value is in the interval } [1, 3]$$

The Mean Value Theorem for Integrals can be extended to find the average function value.

Average Function Value

If a function f has an integral on $[a, b]$, the average value or mean value of f on $[a, b]$ is defined as:

$$f(c) = \frac{1}{b-a}\int_a^b f(x)\,dx$$

1+2=3 MATH TALK!

The average function value comes from the Mean Value Theorem for Integrals if you solve for $f(c)$. The average value of a function is the average height of the graph of a function.

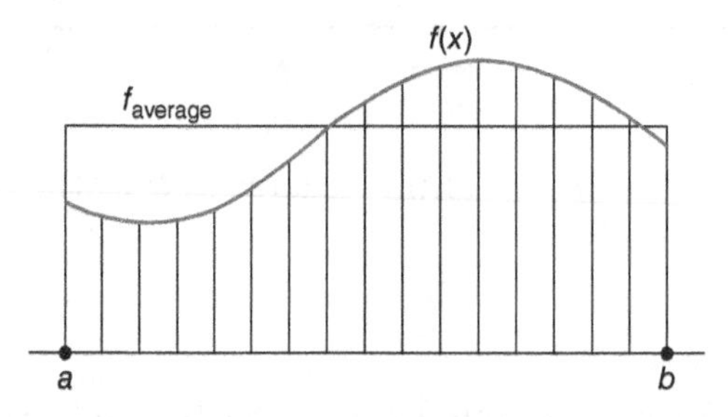

This is the value you find if asked to find the average value of a function.

Example 17:

Find the average value of each of the following functions over the given interval.

1. $f(x) = 3x; [1, 3]$
2. $f(x) = \sin x; [0, \pi]$
3. $f(x) = \sqrt{4 - x^2}; [-2, 2]$

Solution:

1. $f(c) = \frac{1}{3-1}\int_1^3 3x\,dx = \frac{1}{2} \cdot \frac{3x^2}{2}\Big|_1^3 = \frac{3}{4}((3)^2 - (1)^2) = \frac{3}{4} \cdot (8) = 6.$

   The average function value is 6.

2. $f(c) = \frac{1}{\pi - 0}\int_0^\pi \sin x\,dx = \frac{1}{\pi} \cdot -\cos x\Big|_0^\pi = \frac{1}{\pi}(-\cos(\pi) - -\cos(0))$

   $= \frac{1}{\pi}(-(-1) + 1) = \frac{2}{\pi}$. The average function value is $\frac{2}{\pi}$.

3. $f(c) = \frac{1}{2 - -2}\int_{-2}^{2}\sqrt{4 - x^2}\,dx = \frac{1}{4}\int_{-2}^{2}\sqrt{4 - x^2}\,dx$. To integrate, consider the graph of the integrand, and find the area using geometric formulas.

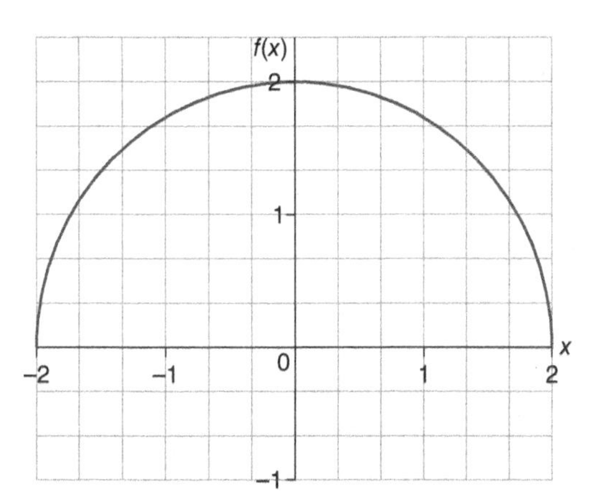

The graph is a semicircle whose area formula is $\frac{\pi r^2}{2}$. The area $= \frac{\pi(2)^2}{2} = 2\pi$:

$$f(c) = \frac{1}{4}(2\pi) = \frac{\pi}{2}$$

Example 18:

The temperature, in degrees Celsius, of the water in a lake is a differentiable function $W(t)$. The table below shows the water temperature as recorded every 4 days over a 20-day period. Approximate the average temperature, in degrees Celsius, of the water over the time interval $0 \leq t \leq 20$ days using a trapezoidal approximation with subintervals of length $\Delta t = 4$ days.

***t* (days)**	0	4	8	12	16	20
***W*(*t*) (°C)**	21	30	26	22	20	23

Solution:

The last sentence states what is being asked for in the problem with the key phrase "average temperature." Since the function represents temperature, the problem is asking for the average function value, $f(c) = \frac{1}{b-a}\int_a^b f(x)\,dx$. To evaluate the integral, the problem states to use a trapezoidal approximation:

$$W(c) = \frac{1}{20-0}\int_0^{20} W(t)$$

$$= \frac{1}{20}\left[\frac{(21+30)\bullet 4}{2} + \frac{(30+26)\bullet 4}{2} + \frac{(26+22)\bullet 4}{2}\right.$$

$$\left. + \frac{(22+20)\bullet 4}{2} + \frac{(20+23)\bullet 4}{2}\right] = 24$$

The average temperature is approximately 24 degrees Celsius over the time interval $0 \leq t \leq 20$ days.

BRAIN TICKLERS Set # 32

1. If $f(x) = x^3$, $a = 2$, $b = 4$, find c guaranteed by the Mean Value Theorem for Integrals.

2. Find the average value of the function $f(x) = 4x$ over the interval [1, 3].

3. Given the graph of $f(x)$ shown below, find the average value of the function over the interval [0, 8].

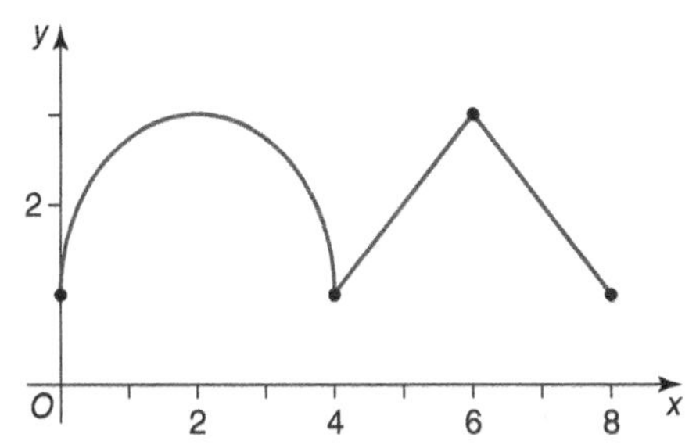

4. As a boiling pot cools, the temperature of the pot is modeled by a differentiable function $P(t)$ for $0 \le t \le 10$, where $P(t)$ is measured in degrees Celsius and time, t, is measured in minutes. Values of $P(t)$ for selected times are shown in the table below. Approximate the average temperature, in degrees Celsius, of the water over the time interval $0 \le t \le 10$ minutes using a left Reimann approximation with subintervals indicated by the table.

***t* (minutes)**	0	2	5	8	10
***P*(*t*) (degrees Celsius)**	65	60	54	48	42

(Answers on page 250.)

BRAIN TICKLERS—THE ANSWERS

Set # 29, pages 217-218

1. 0
2. 9
3. –10
4. 24

Set # 30, page 236

1. 44
2. 18.25
3. –1
4. $e^4 + 2e^2 + 2e^0 + e^{-2} = 2 + e^4 + 2e^2 + e^{-2}$

Set # 31, page 244

1. 1,290
2. $\frac{1}{3}(-e^{-8} + e) = -\frac{1}{3}\left(\frac{1}{e^8} - e\right)$
3. $\frac{e^2}{2} - \frac{1}{2}$
4. $\frac{1}{\sqrt{x^8 + 5}} \cdot 4x^3$

Set # 32, page 249

1. $\sqrt[3]{30}$
2. 8
3. $\frac{1}{8}(2\pi + 4 + 8) = \frac{1}{8}(12 + 2\pi)$
4. 56.8 degrees Celsius

Chapter 7

Applications of Definite Integrals

Knowing how to calculate the definite integral of a function and what it represents leads to some very useful applications of definite integrals.

Area Between Curves

When graphing multiple functions, their graphs may intersect and create regions between the two curves.

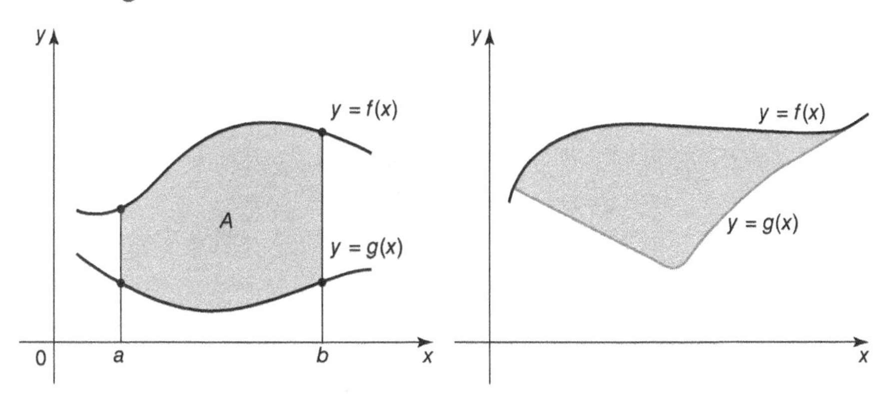

Since the definite integral represents the area under the curve bounded by the curve and the x-axis, the same idea can be applied to the area between two curves.

Consider the graphs of $f(x) = -x^2 + 10$ and $g(x) = x^2 + 2$ shown below. The graphs intersect at $x = -2$ and $x = 2$.

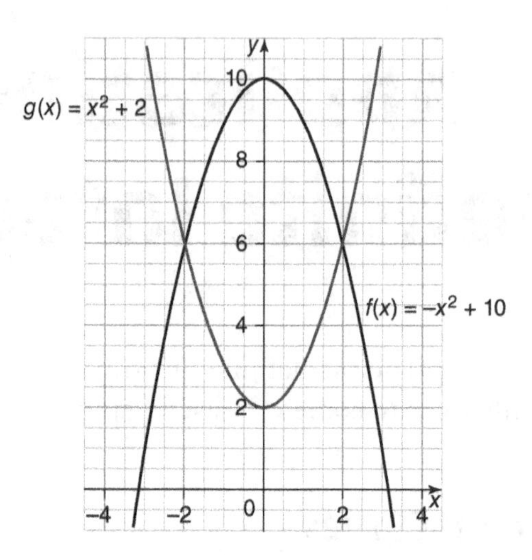

If you find the area under the curve $f(x)$, above the x-axis, and bounded by $x = -2$ and $x = 2$, the graph would be as represented below.

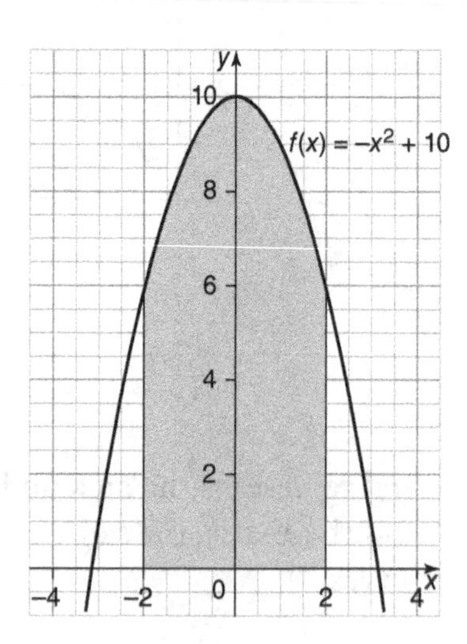

If you find the area under the curve $g(x)$, above the x-axis, and bounded by $x = -2$ and $x = 2$, the graph would be as represented below.

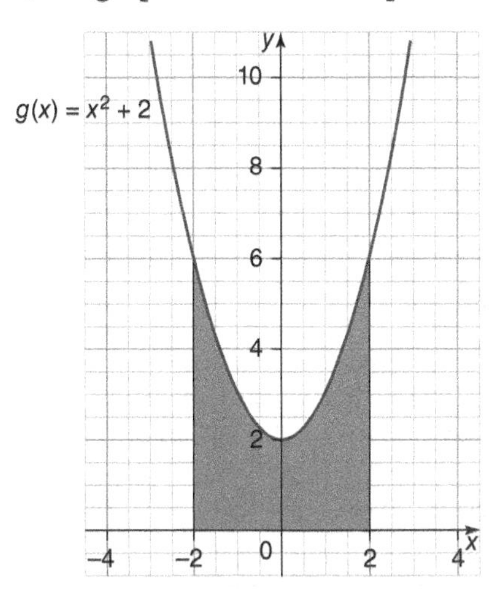

Shown below are the two shaded regions on one graph. To find the area between the two curves, the overlapped areas need to be subtracted away.

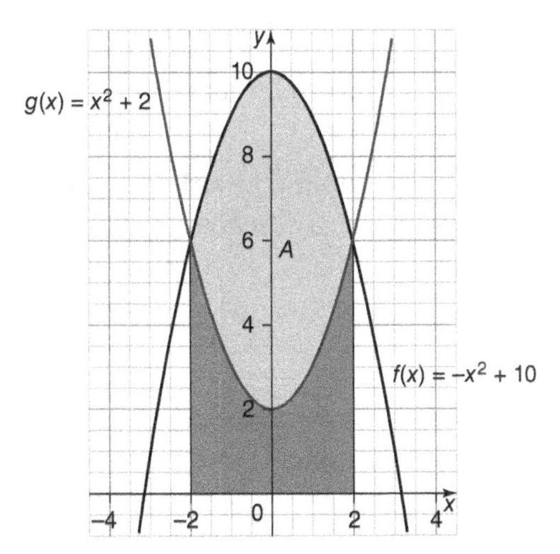

To find the area between the curves $f(x)$ and $g(x)$, calculate the following:

$$A = \int_{-2}^{2} f(x)\,dx - \int_{-2}^{2} g(x)\,dx$$

Using properties of integrals, the above can also be written as:

$$A = \int_{-2}^{2} (f(x) - g(x))\,dx$$

In general, if $f(x)$ and $g(x)$ are continuous on the interval $[a, b]$ and if $f(x) \geq g(x)$ for all x in $[a, b]$, the area between $f(x)$ and $g(x)$ is shown as follows:

$$A = \int_a^b (f(x) - g(x))\, dx = \int_a^b f(x)\, dx - \int_a^b g(x)\, dx$$

1+2=3 MATH TALK!

To find the area between two curves, it is important to see which function is on top and to find the points of intersection between the two curves. Once that is found, the following integral can be evaluated:

$$A = \int_a^b (\text{top function} - \text{bottom function})\, dx$$

Finding the area between two continuous curves is *painless*. It involves the following four steps.

Step 1: Graph the two curves, and determine which one is on top.

Step 2: Find the points of intersection; these will be the boundaries if they are not provided.

Step 3: Write the definite integral as either two separate integrals or one integral as the difference between the top and bottom functions.

Step 4: Evaluate the integral.

Example 1:

Find the area of the region between $y = x^2 + 5$ and $y = x + 1$, from $x = 0$ to $x = 3$.

Solution:

Step 1: Graph the two curves, and determine which one is on top.

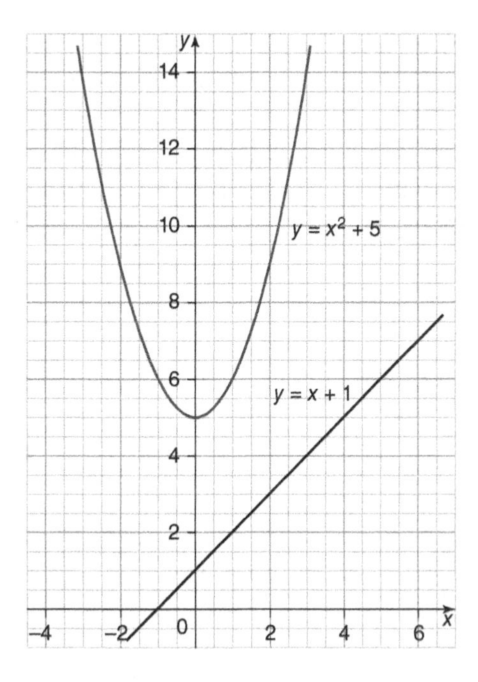

The top function is $y = x^2 + 5$, and the bottom function is $y = x + 1$.

Step 2: Find the points of intersection; these will be the boundaries if they are not provided.

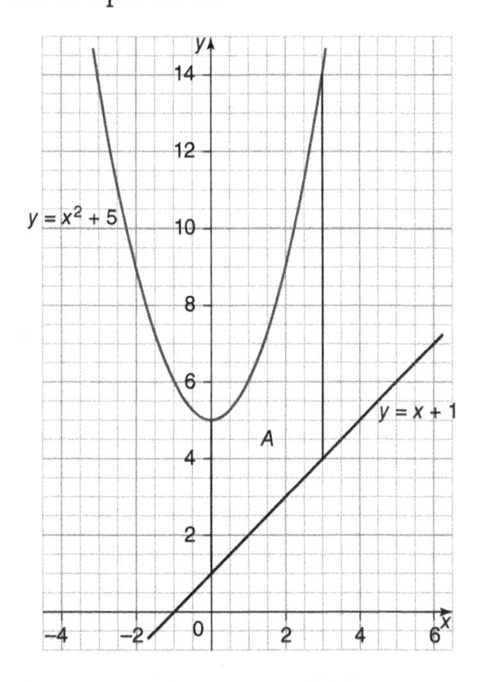

The boundaries of the region are provided and are $x = 0$ and $x = 3$. The top function remains on top over this region.

Step 3: Write the definite integral as either two separate integrals or one integral as the difference between the top and bottom functions:

$$A = \int_0^3 (x^2 + 5 - (x + 1))\, dx = \int_0^3 (x^2 - x + 4)\, dx$$

Step 4: Evaluate the integral:

$$A = \int_0^3 (x^2 - x + 4)\, dx = \left(\frac{x^3}{3} - \frac{x^2}{2} + 4x\right)\Bigg|_0^3$$

$$= \frac{(3)^3}{3} - \frac{(3)^2}{2} + 4(3) - \left(\frac{(0)^3}{3} - \frac{(0)^2}{2} + 4(0)\right)$$

$$A = 16.5$$

CAUTION—Major Mistake Territory!

When approaching a problem about the area between two curves, be sure to keep in mind the following:

1. Functions are continuous and integrable.
2. There are clear points of intersection for the bounds. (These will be given, or you will have to find the points of intersection.)
3. Check that throughout the interval, the top and bottom functions remain the same throughout. In other words, make sure that the top function doesn't become the bottom function at some point. If they change position, then break up the integrals.
4. Do not be concerned about the signs of the integral if the region falls below the x-axis.

Example 2:

Find the area of the region between $y = x$ and $y = -x^2 - 3$ from $x = -1$ to $x = 3$.

Solution:

Step 1: Graph the two curves, and determine which one is on top.

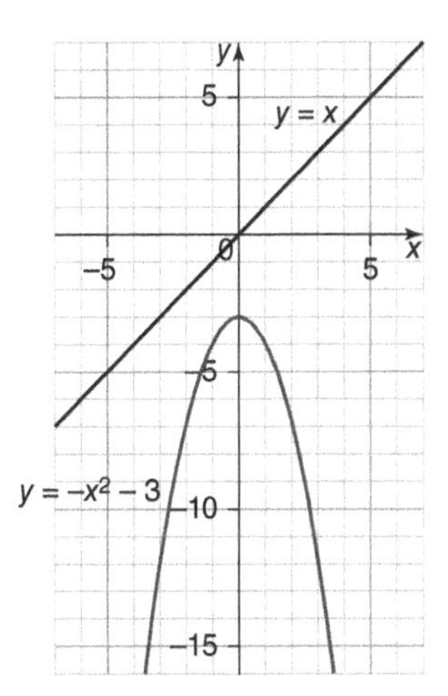

The top function is $y = x$, and the bottom function is $y = -x^2 - 3$.

Step 2: Find the points of intersection; these will be the boundaries if they are not provided.

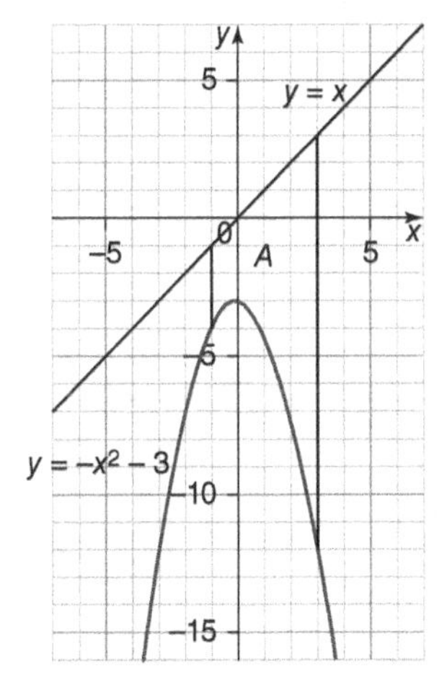

The boundaries of the region are provided and are $x = -1$ to $x = 3$. The top function remains on top over this region.

Step 3: Write the definite integral as either two separate integrals or one integral as the difference between the top and bottom functions:

$$A = \int_{-1}^{3} (x - (-x^2 - 3))\,dx = \int_{-1}^{3} (x^2 + x + 3)\,dx$$

Step 4: Evaluate the integral:

$$A = \int_{-1}^{3} (x^2 + x + 3)\,dx = \left.\frac{x^3}{3} + \frac{x^2}{2} + 3x \right|_{-1}^{3}$$

$$= \frac{(3)^3}{3} + \frac{(3)^2}{2} + 3(3) - \left(\frac{(-1)^3}{3} + \frac{(-1)^2}{2} + 3(-1)\right)$$

$$A = \frac{76}{3}$$

Example 3:

Find the area of the region between $y = -x$ and $y = -x^2 + 2$.

Solution:

Step 1: Graph the two curves, and determine which one is on top.

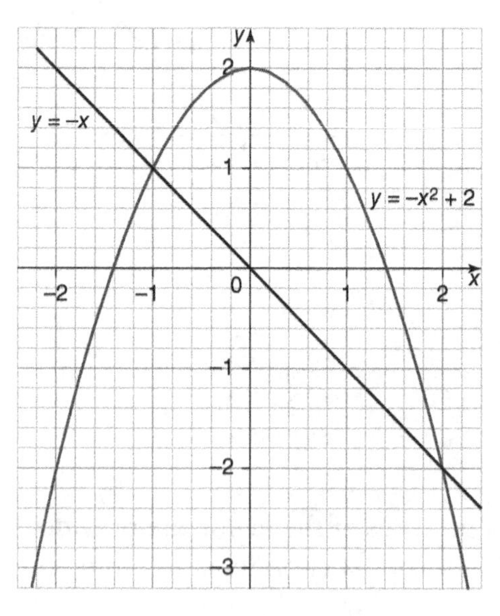

The top function is $y = -x^2 + 2$, and the bottom function is $y = -x$.

Step 2: Find the points of intersection; these will be the boundaries if they are not provided.

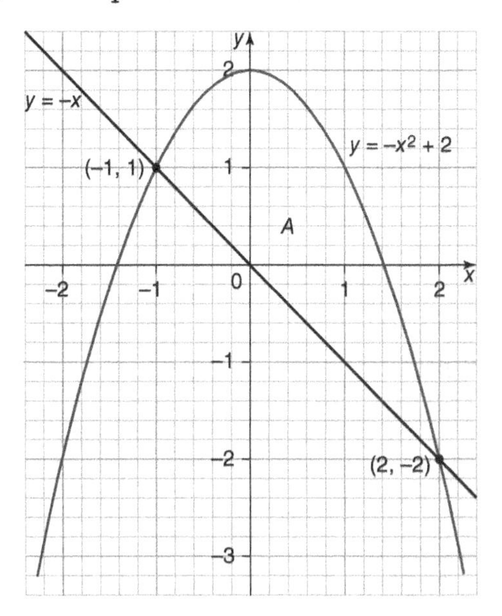

The points of intersection are $(-1, 1)$ and $(2, -2)$. Therefore, the boundaries of the region are $x = -1$ to $x = 2$. The top function remains on top over this region.

Step 3: Write the definite integral as either two separate integrals or one integral as the difference between the top and bottom functions:

$$A = \int_{-1}^{2} (-x^2 + 2 - (-x))\, dx = \int_{-1}^{2} (-x^2 + x + 2)\, dx$$

Step 4: Evaluate the integral:

$$A = \int_{-1}^{2} (-x^2 + x + 2)\, dx = \frac{-x^3}{3} + \frac{x^2}{2} + 2x \Big|_{-1}^{2}$$

$$= \frac{-(2)^3}{3} + \frac{(2)^2}{2} + 2(2) - \left(\frac{-(-1)^3}{3} + \frac{(-1)^2}{2} + 2(-1) \right)$$

$$A = 4.5$$

Example 4:

Find the area of the region bounded by $y^2 = x - 1$ and $y = x - 3$.

Solution:

Step 1: Graph the two curves, and determine which one is on top.

To graph the curve $y^2 = x - 1$, isolate y:

$$y^2 = x - 1 \rightarrow y = +\sqrt{x-1},\ y = -\sqrt{x-1}$$

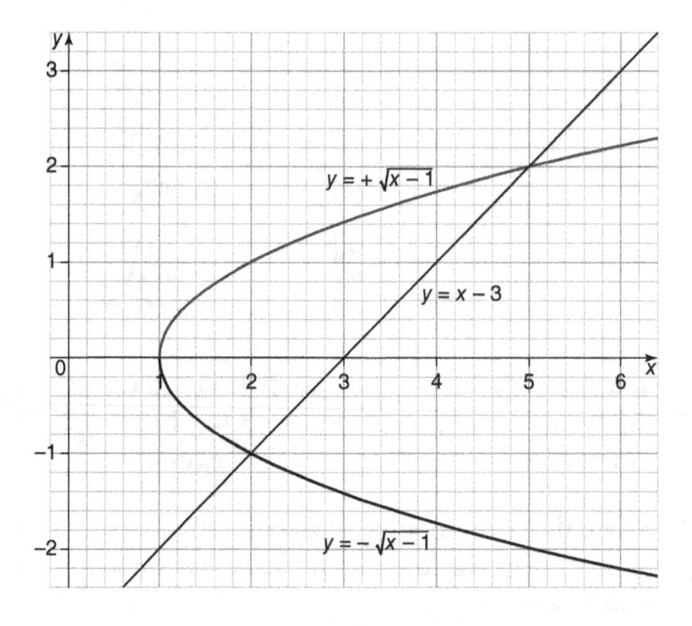

The top function is $y = +\sqrt{x-1}$. However, the bottom function changes over the interval. From $x = 1$ to $x = 2$, the bottom function is $y = -\sqrt{x-1}$.

From $x = 2$ to $x = 5$, the bottom function is $y = x - 3$.

To find the area, the integral will need to be split up for these two different intervals.

Step 2: Find the points of intersection; these will be the boundaries if they are not provided.

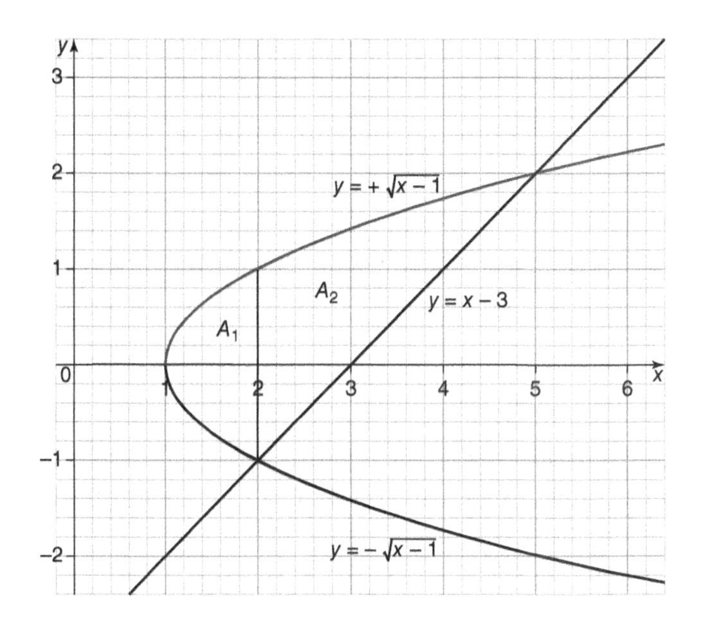

For the first area, A_1, the boundaries of the region are $x = 1$ to $x = 2$.

For the second area, A_2, the boundaries of the region are $x = 2$ to $x = 5$.

Step 3: Write the definite integral as either two separate integrals or one integral as the difference between the top and bottom functions:

$$A = A_1 + A_2$$

$$= \int_1^2 (+\sqrt{x-1} - (-\sqrt{x-1}))\,dx + \int_2^5 (+\sqrt{x-1} - (x-3))\,dx$$

$$= \int_1^2 (2\sqrt{x-1})\,dx + \int_2^5 (\sqrt{x-1} - x + 3)\,dx$$

Step 4: Evaluate the integral:

$$A = 2 \bullet \frac{(x-1)^{3/2}}{\frac{3}{2}} \Bigg|_1^2 + \left(\frac{(x-1)^{3/2}}{\frac{3}{2}} - \frac{x^2}{2} + 3x \right) \Bigg|_2^5$$

$$= \left[\frac{4}{3}((2)-1)^{3/2} - \frac{4}{3}((1)-1)^{3/2} \right]$$

$$+ \left[\left(\frac{2((5)-1)^{3/2}}{3} - \frac{(5)^2}{2} + 3(5) \right) - \left(\frac{2((2)-1)^{3/2}}{3} - \frac{(2)^2}{2} + 3(2) \right) \right]$$

$$A = 4.5$$

There is an alternate way of solving Example 4. Instead of integrating with respect to x where you consider the same top function over the bottom function, you can integrate with respect to y. Then the direction of the function changes, from right to left. In Example 4, the function to the right is always $y = x - 3$, and the function to the left is always $y^2 = x - 1$.

In general, if $f(y)$ and $g(y)$ are continuous on the interval $[c, d]$ and if $f(y) \geq g(y)$ for all y in $[c, d]$, the area between $f(y)$ and $g(y)$ is found with the following:

$$A = \int_c^d (f(y) - g(y))\, dy = \int_c^d f(y)\, dy - \int_c^d g(y)\, dy$$

1+2=3 MATH TALK!

To find the area of two curves, it is important to see which function is on the right. You must find the points of intersection (the lowest point and the highest point) if the boundaries are not provided. Then the following integral can be evaluated:

$$A = \int_c^d (\text{right function} - \text{left function})\, dy$$

It is important to remember when using this approach, everything must be written in terms of y: the boundaries, the functions, and the integral.

A helpful technique to see this new method is to take the graph and rotate it 90° counterclockwise. This way, the top function is the right function and the left function is the bottom function. The boundaries are then the point farthest to the left and the point farthest to the right.

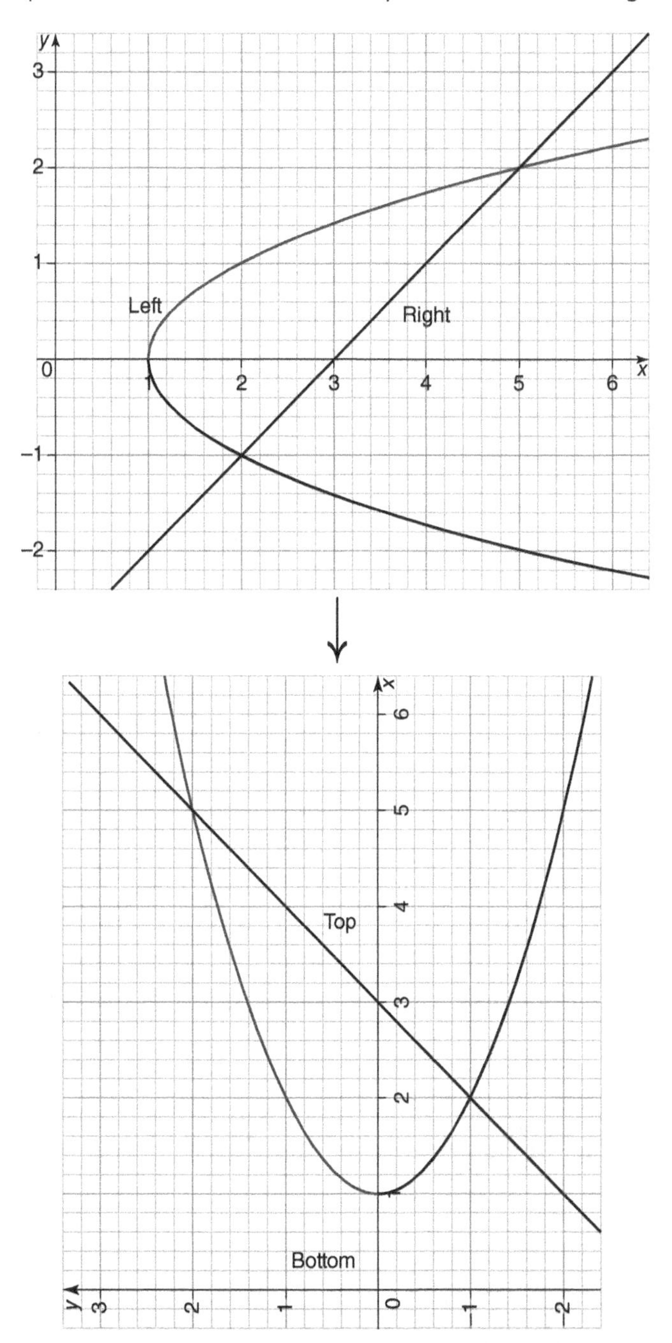

Example 4 revisited:

Find the area of the region bounded by $y^2 = x - 1$ and $y = x - 3$.

Solution:

Step 1: Graph the two curves, and determine which one is on the right.

After graphing, rewrite all the equations in terms of y by solving for x:

$$y^2 = x - 1 \rightarrow x = y^2 + 1$$

$$y = x - 3 \rightarrow x = y + 3$$

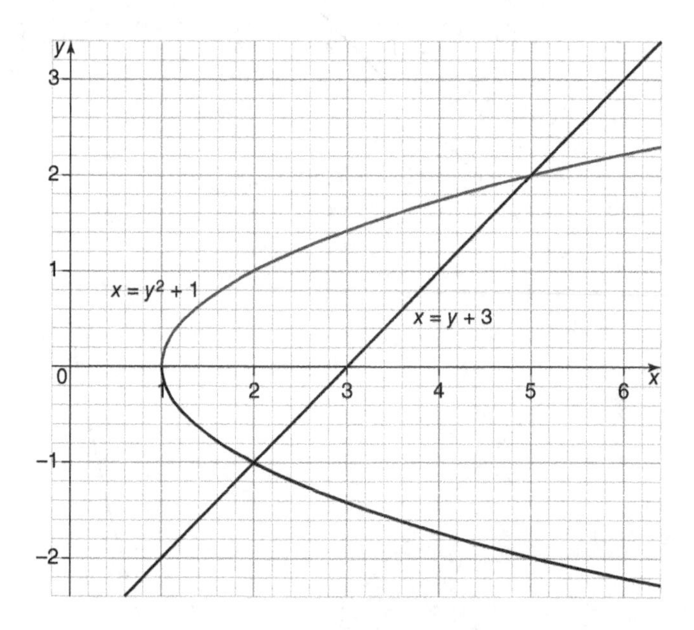

The right function is $x = y + 3$, and the left function is $x = y^2 + 1$.

Step 2: Find the points of intersection; these will be the boundaries if they are not provided.

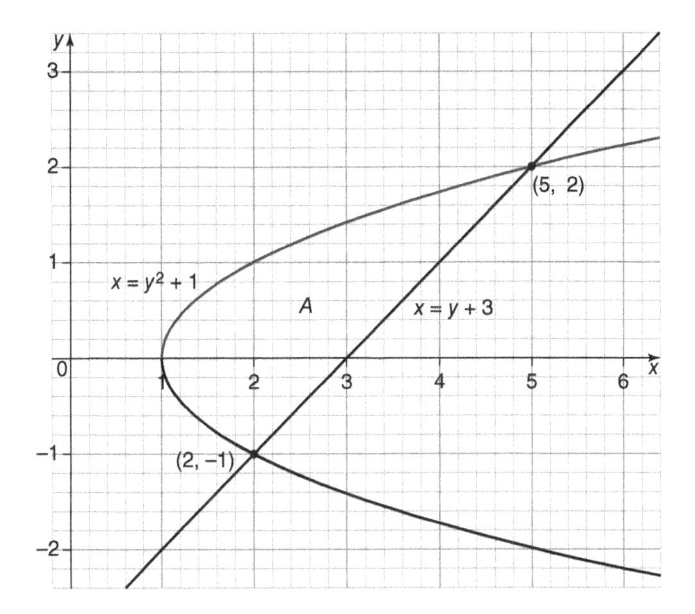

In terms of y, the boundaries are $y = -1$ to $y = 2$.

Step 3: Write the definite integral as either two separate integrals or one integral as the difference between the right and left functions:

$$A = \int_{-1}^{2}(y + 3 - (y^2 + 1))\,dy = \int_{-1}^{2}(-y^2 + y + 2)\,dy$$

Step 4: Evaluate the integral:

$$A = \frac{-y^3}{3} + \frac{y^2}{2} + 2y \Big|_{-1}^{2}$$

$$= \left(\frac{-(2)^3}{3} + \frac{(2)^2}{2} + 2(2)\right) - \left(\frac{-(-1)^3}{3} + \frac{(-1)^2}{2} + 2(-1)\right)$$

$$A = 4.5$$

Both methods resulted in the same solution. This illustrates that it does not matter which method is chosen to evaluate the area problem; the choice is left up to you. Typically, one method will result in fewer integrals to write and evaluate.

Example 5:

Find the area of the region bounded by $x = y^2 + 2$ and the line $x = 11$.

Solution:

Since the equations are in terms of y, you will have to write fewer integrals if you consider the right and left functions instead of the top and bottom functions.

Step 1: Graph the two curves, and determine which one is on the right.

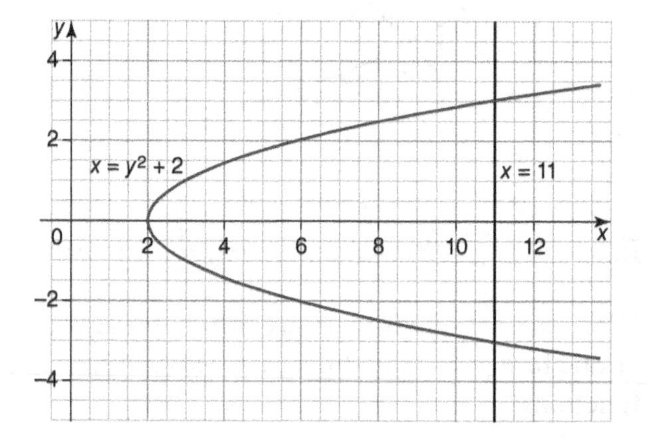

The right curve is $x = 11$, and the left curve is $x = y^2 + 2$.

Step 2: Find the points of intersection; these will be the boundaries if they are not provided.

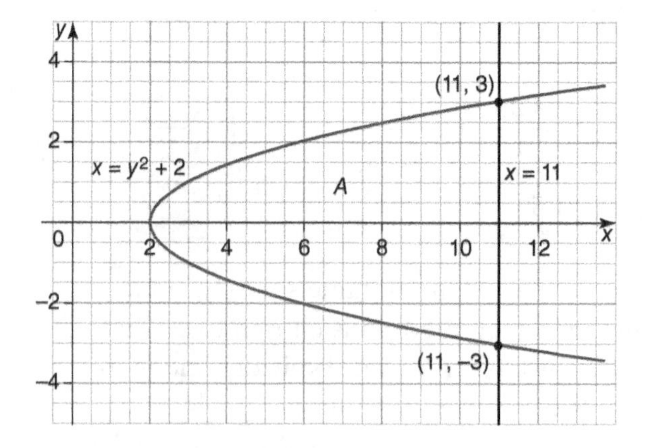

In terms of y, the boundaries are $y = -3$ to $y = 3$.

Step 3: Write the definite integral as either two separate integrals or one integral as the difference between the right and left functions:

$$A = \int_{-3}^{3} (11 - (y^2 + 2))\, dy = \int_{-3}^{3} (-y^2 + 9)\, dy$$

Step 4: Evaluate the integral:

$$A = \frac{-y^3}{3} + 9y \Big|_{-3}^{3}$$

$$= \left(\frac{-(3)^3}{3} + 9(3)\right) - \left(\frac{-(-3)^3}{3} + 9(-3)\right)$$

$$A = 36$$

BRAIN TICKLERS Set # 33

1. Find the area of the region between $y = -x^2 - 1$ and $y = x + 5$, from $x = 0$ to $x = 3$.
2. Find the area of the region between $y = x$ and $y = x^2$.
3. Find the area of A, the region in the first quadrant bounded by the graph of $y = 2\sqrt{x}$, the horizontal line $y = 6$, and the y-axis.
4. Find the area of the region bounded by the curves $x = y^2$ and $x = 4$.

(Answers are on page 298.)

Volume of a Solid of Revolution

In geometry courses, common shapes such as a semicircle, rectangle, and triangle are revolved around a side to create three-dimensional figures like the sphere, cylinder, and cone.

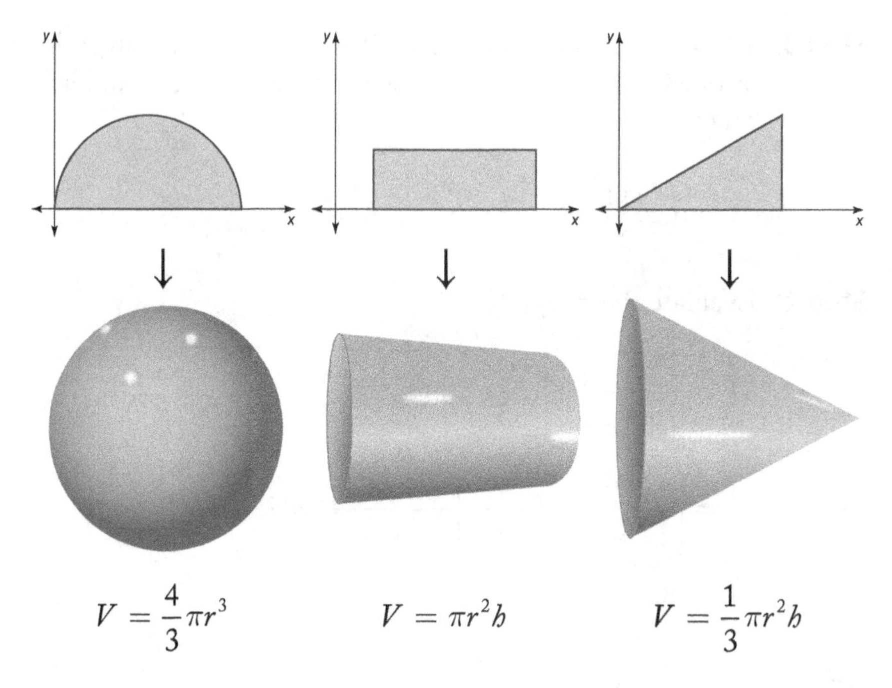

These three-dimensional figures come with their own volume formulas as shown above.

The definite integral can be used to find volumes of solids revolved around an axis. One method is known as the Disc Method. The method is called this because if you slice the three-dimensional model in such a way that it creates discs, you can find the volumes of those discs using the cylinder formula.

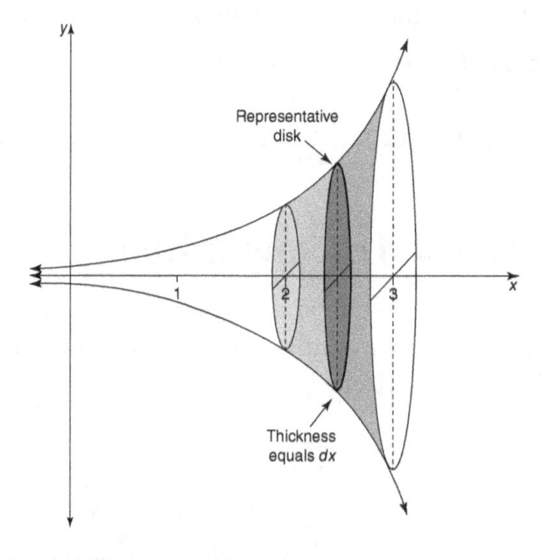

If all the volumes of the discs that make up the figure are added up, you have found the volume of the figure. This can be simplified by using the following integral:

$$V = \int_a^b \pi(f(x))^2\,dx = \pi\int_a^b (f(x))^2\,dx$$

This integral means that the function is revolved around the x-axis.

If the axis of revolution is the y-axis, the integral changes to the following:

$$V = \int_c^d \pi(f(y))^2\,dy = \pi\int_c^d (f(y))^2\,dy$$

1+2=3 **MATH TALK!**

The volume integral is very similar to the volume formula for a cylinder, which makes sense since the discs are like cylinders.

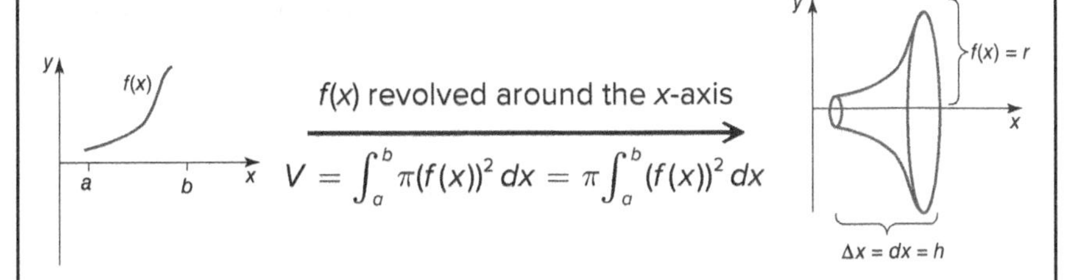

The integral above acts as a summation, where $f(x)$ represents the radius and the dx represents the height.

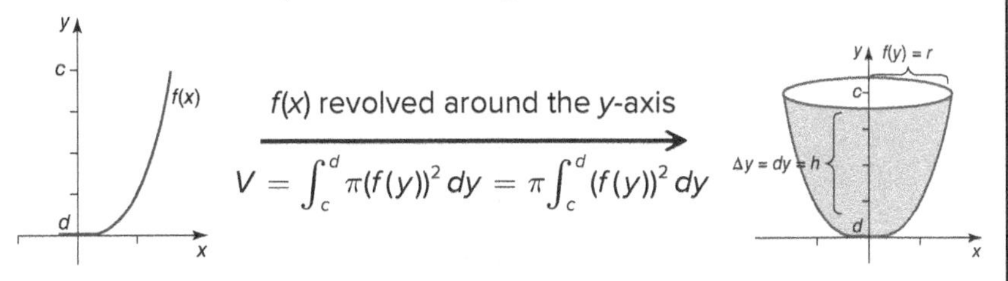

The integral above acts as a summation, where $f(y)$ represents the radius and the dy represents the height.

Finding the volume of solids of revolution is *painless*. It involves the following four steps.

Step 1: Graph the function, and identify if the axis of revolution is horizontal or vertical. This will determine which variable the interval will be written with respect to.

Step 2: Find the points of intersection; these will be the boundaries if they are not provided.

Step 3: Write the definite integral.

Step 4: Evaluate the integral.

Example 6:

Find the volume of the solid generated when the region between the graph of the equation $y = x^2$ from $x = 2$ to $x = 5$ is revolved about the x-axis.

Solution:

Step 1: Graph the function, and identify if the axis of revolution is horizontal or vertical. This will determine which variable the interval will be written with respect to.

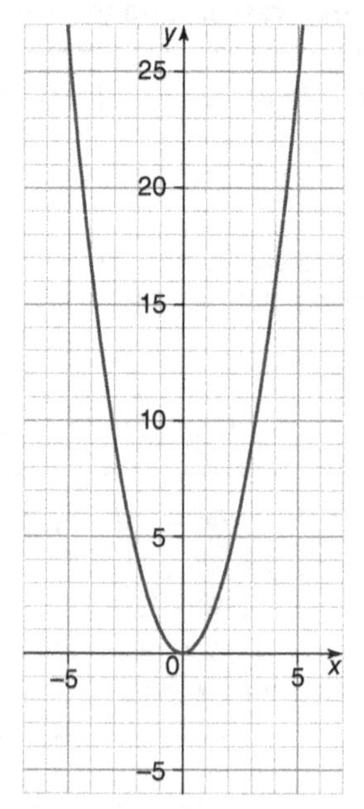

The axis of revolution is the x-axis, which is horizontal. Therefore, the integral will be written with respect to x.

Step 2: Find the points of intersection; these will be the boundaries if they are not provided.

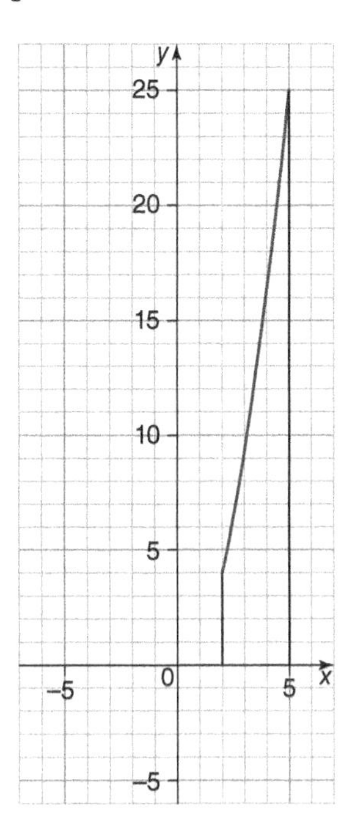

The boundaries were provided and are $x = 2$ and $x = 5$.

Step 3: Write the definite integral:

$$V = \pi \int_2^5 (x^2)^2 \, dx$$

Step 4: Evaluate the integral:

$$V = \pi \int_2^5 x^4 \, dx = \pi \bullet \frac{x^5}{5} \Bigg|_2^5 = \pi \left(\frac{(5)^5}{5} - \frac{(2)^5}{5} \right)$$

$$= \pi \left(\frac{3125}{5} - \frac{32}{5} \right)$$

$$V = \frac{3093}{5} \pi$$

Example 7:

Find the volume of the solid generated when the region, in Quadrant I, between the graphs of the equations $y = x^2$, $y = 1$ and $x = 0$ is revolved about the y-axis.

Solution:

Step 1: Graph the function, and identify if the axis of revolution is horizontal or vertical. This will determine which variable the interval will be written with respect to.

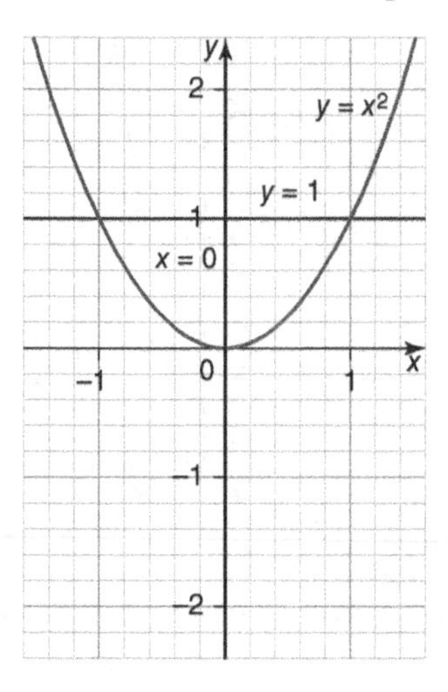

The axis of revolution is the y-axis, which is vertical. Therefore, the integral will be written with respect to y. The equation written with respect to y in the first quadrant is

$$y = x^2 \rightarrow x = +\sqrt{y}$$

Step 2: Find the points of intersection; these will be the boundaries if they are not provided.

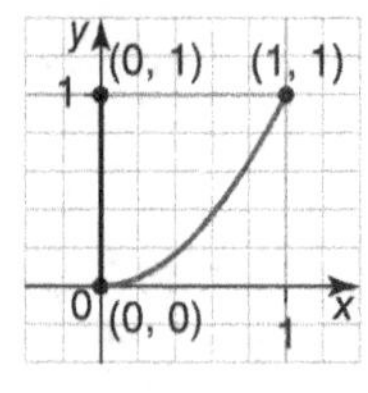

The boundaries are found at the points of intersection and are $y = 0$ and $y = 1$.

Step 3: Write the definite integral:

$$V = \pi\int_0^1 (\sqrt{y})^2\, dy$$

Step 4: Evaluate the integral:

$$V = \pi\int_0^1 y\, dy = \pi \bullet \frac{y^2}{2}\bigg|_0^1 = \pi\left(\frac{(1)^2}{2} - \frac{(0)^2}{2}\right) = \pi\left(\frac{1}{2} - 0\right)$$

$$V = \frac{1}{2}\pi$$

Like with area, you can rotate a graph that is made up of two curves around an axis to create a solid. To calculate the volume, we use what is called the Washer Method. Much like a washer in real life, it is circular with an outer curve and has an inner curve that forms a hole in the middle.

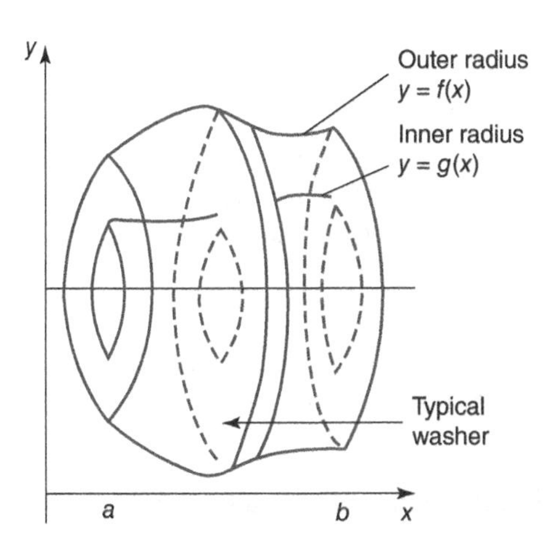

Like finding the area between two curves, to find the volume using the Washer Method by rotating around the x-axis (as shown above), use the following integral:

$$V = \pi\int_a^b (f(x))^2\, dx - \pi\int_a^b (g(x))^2\, dx = \pi\int_a^b (f(x))^2 - (g(x))^2\, dx$$

To find the volume using the Washer Method by rotating around the y-axis (as shown below) use the following integral:

$$V = \pi\int_c^d (f(y))^2\,dy - \pi\int_c^d (g(y))^2\,dy = \pi\int_c^d (f(y))^2 - (g(y))^2\,dy$$

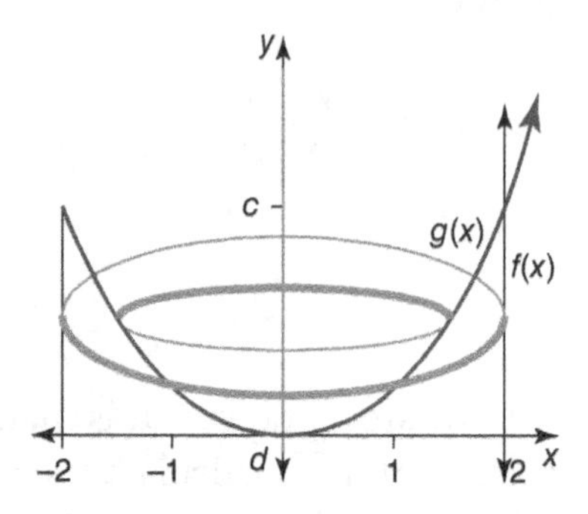

1+2=3 MATH TALK!

To find the volume of a solid of revolution formed by two curves, you must know if the solid is revolved around a horizontal axis (like the x-axis) or a vertical axis (like the y-axis).

If the functions are revolved around a horizontal axis, you must note which function is on top and find the points of intersection between the two curves. Once that has been found, evaluate the following integral:

$$V = \pi\int_a^b ((\text{top function})^2 - (\text{bottom function})^2)\,dx$$

If the functions are revolved around a vertical axis, you must note which function is on the right and find the points of intersection between the two curves. Once that has been found, evaluate the following integral:

$$V = \pi\int_c^d ((\text{right function})^2 - (\text{left function})^2)\,dy$$

Example 8:

Find the volume of the solid generated when the region between the graphs of the equations $y = x^2$ and $y = 2x$ is revolved about the x-axis.

Solution:

Step 1: Graph the function, and identify if the axis of revolution is horizontal or vertical. This will determine which variable the interval will be written with respect to.

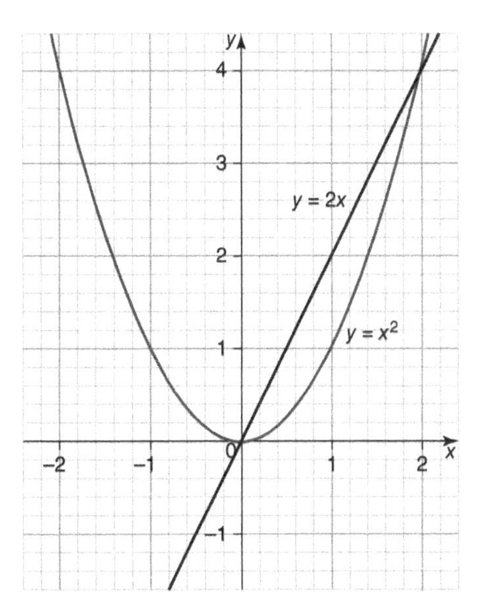

The axis of revolution is the x-axis, which is horizontal. Therefore, the integral will be written with respect to x. The top function is $y = 2x$, and the bottom function is $y = x^2$.

Step 2: Find the points of intersection; these will be the boundaries if they are not provided.

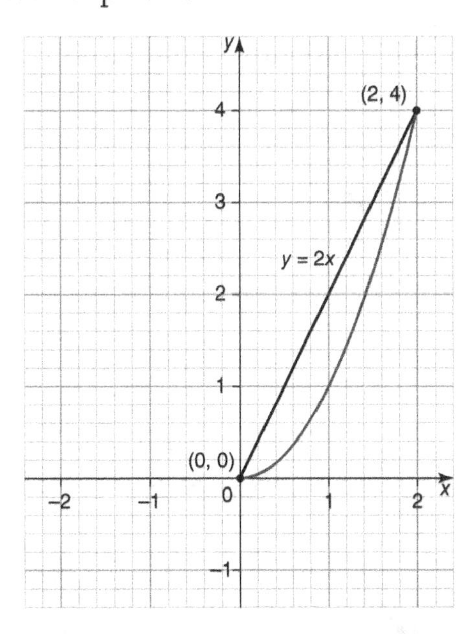

The boundaries are found at the points of intersection and are $x = 0$ and $x = 2$.

Step 3: Write the definite integral:

$$V = \pi\int_0^2 (2x)^2\,dx - \pi\int_0^2 (x^2)^2\,dx = \pi\int_0^2 ((2x)^2 - (x^2)^2)\,dx$$

Step 4: Evaluate the integral:

$$V = \pi\int_0^2 (4x^2 - x^4)\,dx = \pi\left(\frac{4x^3}{3} - \frac{x^5}{5}\right)\Bigg|_0^2$$

$$= \pi\left[\left(\frac{4(2)^3}{3} - \frac{(2)^5}{5}\right) - \left(\frac{4(0)^3}{3} - \frac{(0)^5}{5}\right)\right] = \pi\left[\left(\frac{32}{3} - \frac{32}{5}\right) - (0 - 0)\right]$$

$$V = \frac{64}{15}\pi$$

Example 9:

Find the volume of the solid generated when the region between the graphs of the equations $y = x^2$ and $y = 4$ is revolved about $y = -1$.

Solution:

Step 1: Graph the function, and identify if the axis of revolution is horizontal or vertical. This will determine which variable the interval will be written with respect to.

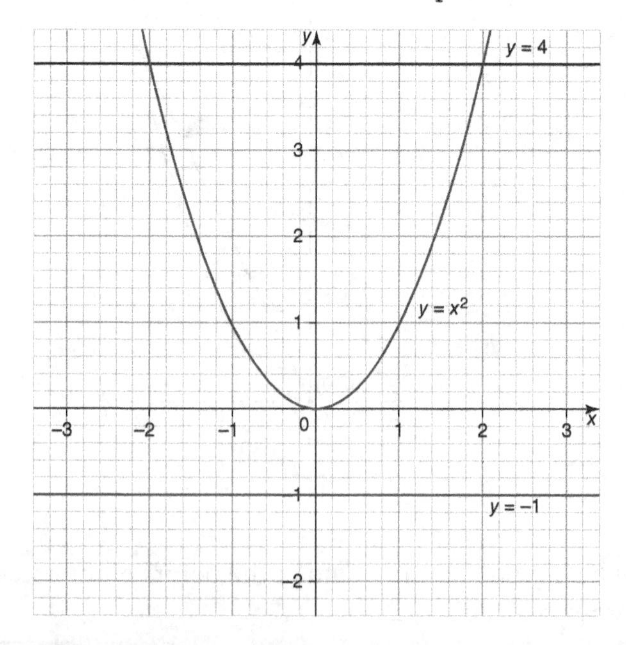

The axis of revolution is the line $y = -1$. Since we are used to revolving around the axes, the given axis of revolution can be shifted up 1 unit so it can become the x-axis. To do this, every equation needs to be shifted up by 1. The new equations will be

$$y = 4 \rightarrow y = 5$$

$$y = x^2 \rightarrow y = x^2 + 1$$

$$y = -1 \rightarrow y = 0, x\text{-axis}$$

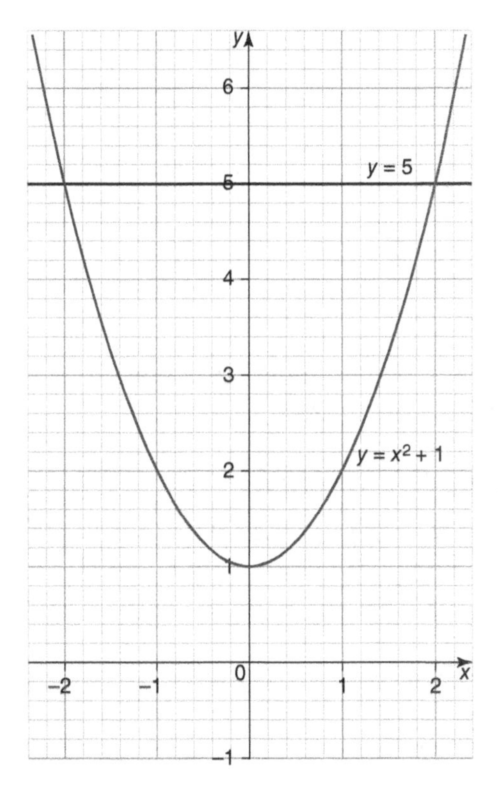

The axis of revolution is the x-axis. Therefore, the integral will be with respect to x. The top function is $y = 5$, and the bottom function is $y = x^2 + 1$.

Step 2: Find the points of intersection; these will be the boundaries if they are not provided.

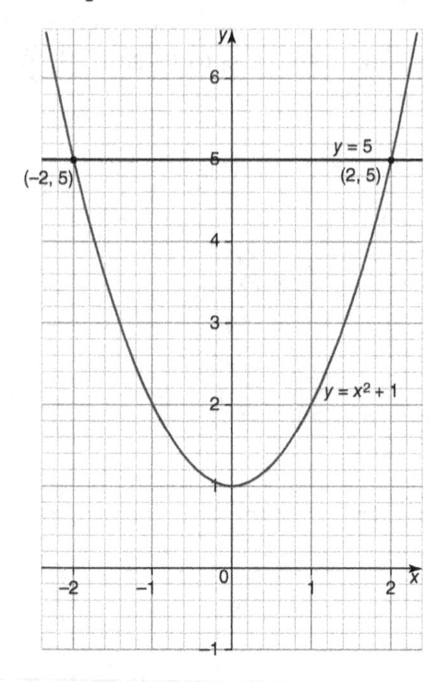

The boundaries are found at the points of intersection and are $x = -2$ and $x = 2$.

Step 3: Write the definite integral:

$$V = \pi \int_{-2}^{2} (5)^2\, dx - \pi \int_{-2}^{2} (x^2 + 1)^2\, dx = \pi \int_{-2}^{2} ((5)^2 - (x^2 + 1)^2)\, dx$$

Step 4: Evaluate the integral:

$$V = \pi\left(\int_{-2}^{2} 25 - (x^4 + 2x^2 + 1)\right) dx = \pi \int_{-2}^{2} (-x^4 - 2x^2 + 24)\, dx$$

$$= \pi\left(\frac{-x^5}{5} - \frac{2x^3}{3} + 24x\right)\Bigg|_{-2}^{2}$$

$$= \pi\left[\left(\frac{-(2)^5}{5} - \frac{2(2)^3}{3} + 24(2)\right) - \left(\frac{-(-2)^5}{5} - \frac{2(-2)^3}{3} + 24(-2)\right)\right]$$

$$= \pi\left[\left(\frac{-32}{5} - \frac{16}{3} + 48\right) - \left(\frac{32}{5} + \frac{16}{3} - 48\right)\right]$$

$$V = \frac{1088\pi}{15}$$

BRAIN TICKLERS Set # 34

1. Find the volume of the solid generated when the region between the graph of the equation $y = 2x + 1$ from $x = 1$ to $x = 3$ is revolved about the x-axis.
2. Find the volume of the solid generated by the region, in the first quadrant, bounded by the graph of $y = \sqrt{x}$, the horizontal line $y = 4$, and the y-axis when it is revolved about the y-axis.
3. Find the volume of the solid generated by revolving the region enclosed by the graphs of $y = x^3$ and $y = \sqrt{x}$ about the x-axis.
4. Find the volume of the solid generated by the region, in the first quadrant, enclosed by the graph of $y = x^2$, the line $x = 2$, and the x-axis when it is revolved about the y-axis.

(Answers on page 298.)

Volume of a Solid with a Known Cross Section

Not all three-dimensional figures have a circular base, because rotating around an axis is not the only way to create a solid. Like pyramids or a cube, three-dimensional figures can also be created using perpendicular cross sections.

These types of figures are generated by a given function forming a region with the axes or other functions and boundaries. From this region, other shapes rise perpendicular to the base, forming the other sides of the solid.

In the figure below, the base is formed by the function $y = 1 - x^2$ and the x-axis. Its cross sections are equilateral triangles perpendicular to the x-axis.

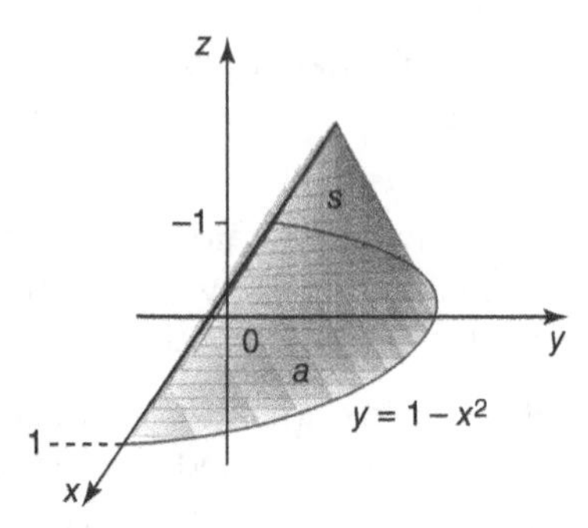

To find the volume of this type of solid, use the following integral:

$$V = \int_a^b A(x)\,dx$$

$A(x)$ is the area function of the base of the solid, which is the shape of the cross section.

Finding the volume of a solid formed by perpendicular cross sections is *painless*. It involves the following four steps.

Step 1: Graph the function that forms the base.

Step 2: Identify the area formula for the shape of the perpendicular cross section.

Step 3: Write the area formula as a function of either x or y, whichever axis the shape is perpendicular to.

Step 4: Evaluate the volume integral.

1+2=3 MATH TALK!

The typical base shapes that are used are the square, rectangle, semicircle, and equilateral triangle. The area formulas for each are listed below:

$$\text{Area of Square} = (\text{side})^2$$

$$\text{Area of Rectangle} = (\text{base})(\text{height})$$

$$\text{Area of Semicircle} = \frac{\pi(\text{radius})^2}{2}$$

$$\text{Area of Equilateral Triangle} = \frac{(\text{side})^2\sqrt{3}}{4}$$

Example 10:

Find the volume of the solid whose base is $y = x + 1$ from $x = 0$ to $x = 3$ and whose cross sections perpendicular to the x-axis are squares.

Solution:

Step 1: Graph the function that forms the base.

The base is formed by the equation $y = x + 1$ from $x = 0$ to $x = 3$.

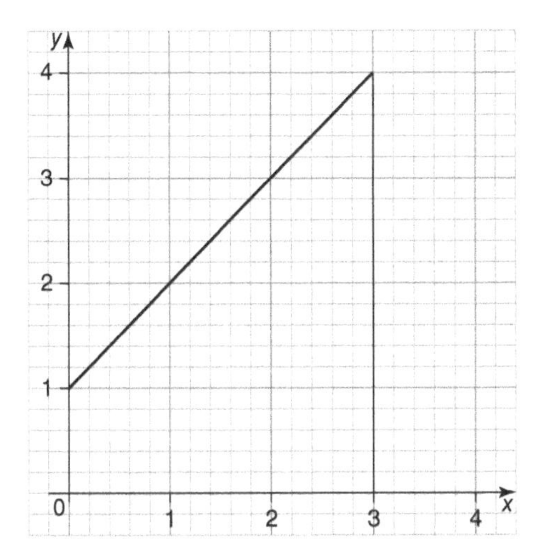

Step 2: Identify the area formula for the shape of the perpendicular cross section.

The shape of the cross section is a square. The area formula for a square is $(\text{side})^2$.

Step 3: Write the area formula as a function of either x or y, whichever axis the shape is perpendicular to.

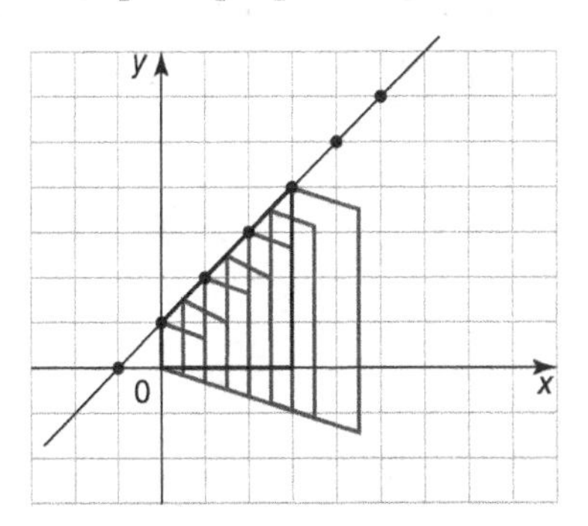

The sides of each square start at the x-axis, go up perpendicular to the x-axis, and stop when they reach the function. Therefore, each side is determined by the function $f(x) = x + 1$. The area formula in terms of x will be the following:

$$A(x) = (x + 1)^2$$

Step 4: Evaluate the volume integral:

$$V = \int_0^3 (x + 1)^2\, dx$$

Let $u = x + 1$, $du = 1dx$

$$\int_0^3 (x + 1)^2\, dx \rightarrow \int_{x=0}^{x=3} (u)^2\, du = \frac{u^3}{3}\Bigg|_{x=0}^{x=3} = \frac{(x+1)^3}{3}\Bigg|_0^3$$

$$= \frac{(3+1)^3}{3} - \frac{(0+1)^3}{3}$$

$$V = \frac{(4)^3}{3} - \frac{(1)^3}{3} = \frac{63}{3} = 21$$

Example 11:

Find the volume of the solid whose base is $y = \sqrt{x}$ from $x = 0$ to $x = 9$ and whose cross sections perpendicular to the x-axis are rectangles with a height of 2.

Solution:

Step 1: Graph the function that forms the base.

The base is formed by the equation $y = \sqrt{x}$ from $x = 0$ to $x = 9$.

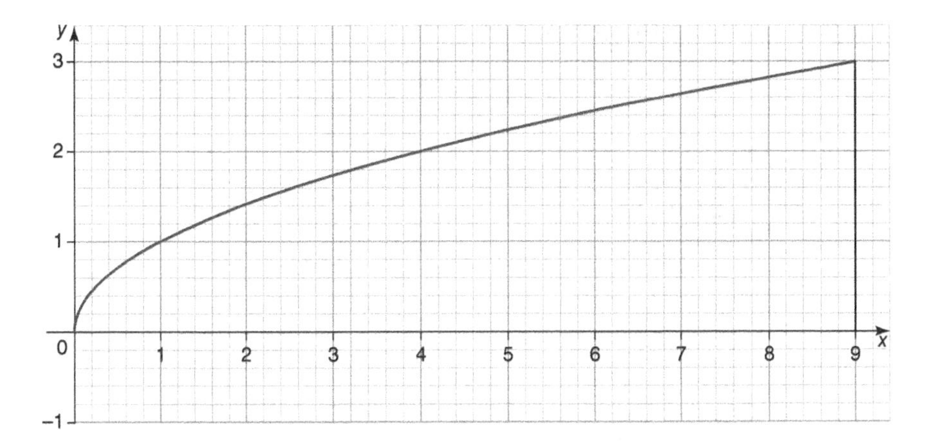

Step 2: Identify the area formula for the shape of the perpendicular cross section.

The shape of the cross section is a rectangle. The area formula for a rectangle is (base)(height).

Step 3: Write the area formula as a function of either x or y, whichever axis the shape is perpendicular to.

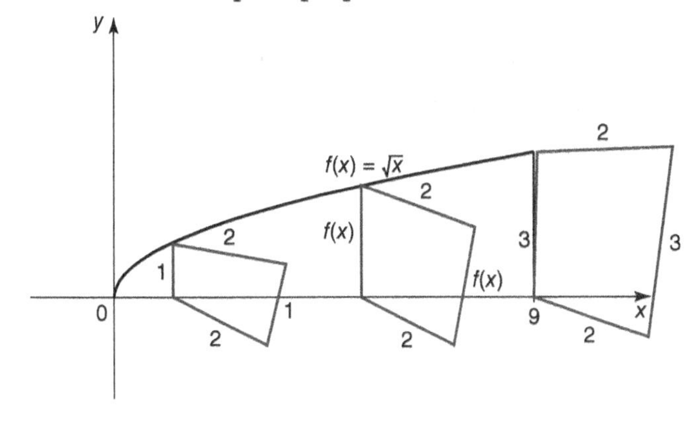

One side of the rectangle starts at the x-axis, goes up perpendicular to the x-axis, and stops when it reaches the function. Therefore, one side is determined by the function $f(x) = \sqrt{x}$. The other side of the rectangle is given as 2. The area formula in terms of x will be the following:

$$A(x) = (2)(\sqrt{x}) = 2\sqrt{x}$$

Step 4: Evaluate the volume integral:

$$V = \int_0^9 (2\sqrt{x})\,dx$$

$$\int_0^9 (2)(\sqrt{x})\,dx = 2\int_0^9 x^{1/2}\,dx = 2 \bullet \frac{x^{3/2}}{\frac{3}{2}}\Bigg|_0^9 = \frac{4x^{3/2}}{3}\Bigg|_0^9$$

$$= \frac{4(9)^{3/2}}{3} - \frac{4(0)^{3/2}}{3}$$

$$V = \frac{108}{3} - \frac{0}{3} = \frac{108}{3} = 36$$

Example 12:

Find the volume of the solid whose base is $y = \sqrt{x}$ from $x = 0$ to $x = 9$ and whose cross sections perpendicular to the x-axis are semicircles.

Solution:

Step 1: Graph the function that forms the base.

The base is formed by the equation $y = \sqrt{x}$ from $x = 0$ to $x = 9$.

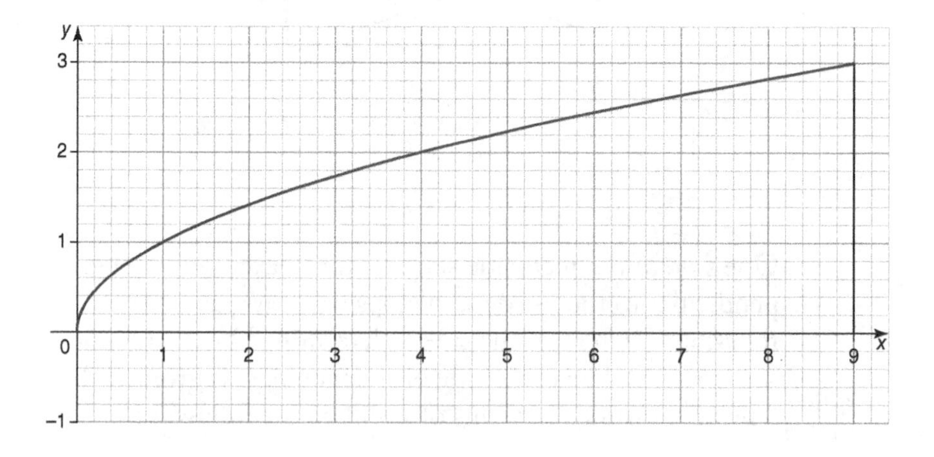

Step 2: Identify the area formula for the shape of the perpendicular cross section.

The shape of the cross section is a semicircle. The area formula for a semicircle is $\frac{\pi r^2}{2}$.

Step 3: Write the area formula as a function of either x or y, whichever axis the shape is perpendicular to.

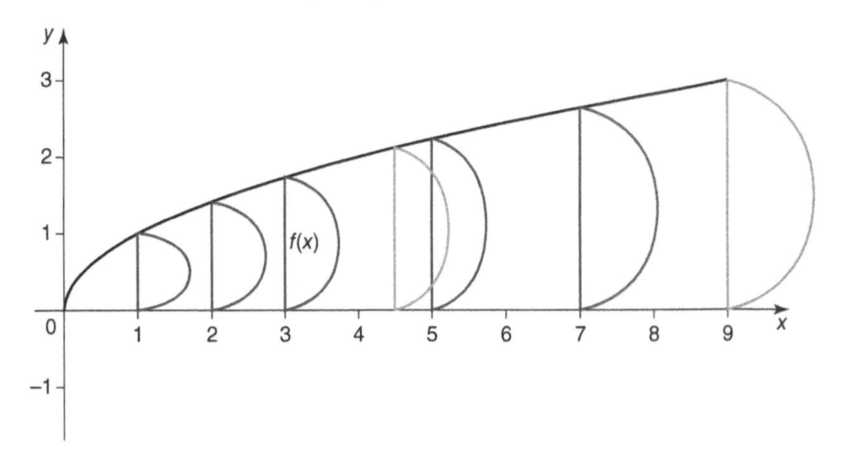

The diameter of the semicircle starts at the x-axis, goes up perpendicular to the x-axis, and stops when it reaches the function. Therefore, the diameter is determined by the function $f(x) = \sqrt{x}$ and the radius $= \frac{\sqrt{x}}{2}$. The area formula in terms of x will be the following:

$$A(x) = \frac{\pi\left(\frac{\sqrt{x}}{2}\right)^2}{2} = \frac{\pi \bullet \frac{x}{4}}{2} = \frac{\pi x}{8}$$

Step 4: Evaluate the volume integral:

$$V = \int_0^9 \frac{\pi x}{8}\, dx$$

$$\int_0^9 \frac{\pi x}{8}\, dx = \frac{\pi}{8}\int_0^9 x\, dx = \frac{\pi}{8} \bullet \frac{x^2}{2}\bigg|_0^9 = \frac{\pi x^2}{16}\bigg|_0^9 = \frac{\pi(9)^2}{16} - \frac{\pi(0)^2}{16}$$

$$V = \frac{81\pi}{16} - \frac{0}{16} = \frac{81\pi}{16}$$

Example 13:

Find the volume of the solid whose base is $y = x^2$ from $y = 0$ to $y = 4$, in Quadrant I, and whose cross sections perpendicular to the y-axis are squares.

Solution:

Step 1: Graph the function that forms the base.

The base is formed by the equation $y = x^2$ from $y = 0$ to $y = 4$.

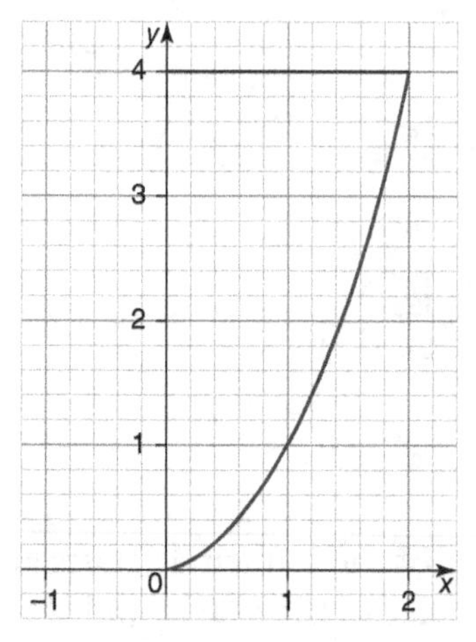

Step 2: Identify the area formula for the shape of the perpendicular cross section.

The shape of the cross section is a square. The area formula for a square is $(\text{side})^2$.

Step 3: Write the area formula as a function of either x or y, whichever axis the shape is perpendicular to.

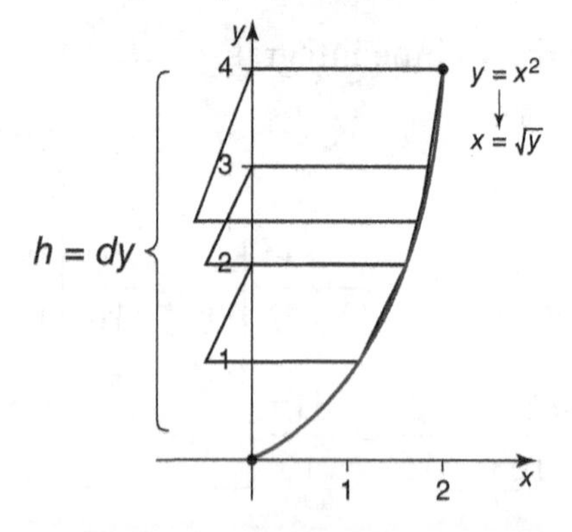

The side of the square starts at the y-axis, goes to the right perpendicular to the y-axis, and stops when it reaches the function. Therefore, the side is determined by the function $f(x) = x^2$ but written in terms of y, $f(y) = \sqrt{y}$. The area formula in terms of y will be the following:

$$A(x) = (\sqrt{y})^2 = y$$

Step 4: Evaluate the volume integral:

$$V = \int_0^4 y\,dy$$

$$\int_0^4 y\,dy = \frac{y^2}{2}\bigg|_0^4 = \frac{(4)^2}{2} - \frac{(0)^2}{2}$$

$$V = \frac{16}{2} - \frac{0}{2} = \frac{16}{2} = 8$$

Example 14:

Find the volume of the solid whose base is the region between $y = x^2$ and $y = x$ and whose cross sections perpendicular to the x-axis are squares.

Solution:

Step 1: Graph the function that forms the base.

The base is formed by the equations $y = x^2$ and $y = x$.

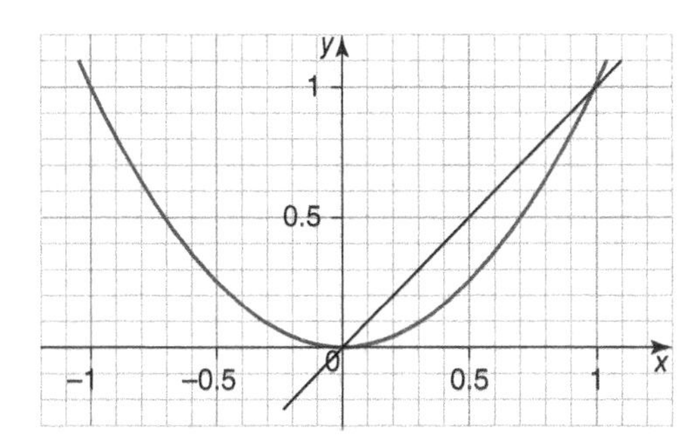

The boundaries of the region are determined by the points of intersection, (0, 0) and (1, 1).

The base is formed by the equations $y = x^2$ and $y = x$ from $x = 0$ to $x = 1$.

Step 2: Identify the area formula for the shape of the perpendicular cross section.

The shape of the cross section is a square. The area formula for a square is $(\text{side})^2$.

Step 3: Write the area formula as a function of either x or y, whichever axis the shape is perpendicular to.

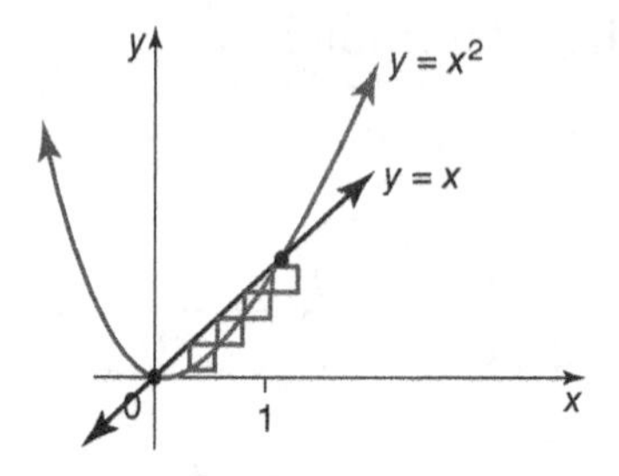

One side of the square starts at the function $y = x^2$, goes up perpendicular to the x-axis, and stops when it reaches the function $y = x$. Therefore, the side is determined by the difference between the top function and the bottom function, $x - x^2$. The area formula in terms of x will be the following:

$$A(x) = (x - x^2)^2 = x^2 - 2x^3 + x^4$$

Step 4: Evaluate the volume integral:

$$V = \int_0^1 (x^2 - 2x^3 + x^4)\,dx$$

$$\int_0^1 (x^2 - 2x^3 + x^4)\,dx = \frac{x^3}{3} - \frac{2x^4}{4} + \frac{x^5}{5}\bigg|_0^1$$

$$= \frac{(1)^3}{3} - \frac{2(1)^4}{4} + \frac{(1)^5}{5} - \left(\frac{(0)^3}{3} - \frac{2(0)^4}{4} + \frac{(0)^5}{5}\right)$$

$$V = \frac{1}{3} - \frac{2}{4} + \frac{1}{5} - \frac{0}{3} + \frac{0}{4} - \frac{0}{5} = \frac{1}{30}$$

BRAIN TICKLERS Set # 35

1. Find the volume of the solid whose base is $y = \sqrt{x}$ from $x = 0$ to $x = 9$ and whose cross sections perpendicular to the x-axis are squares.
2. Find the volume of the solid whose base is $y = x^2$ from $y = 0$ to $y = 4$, in Quadrant I, and whose cross sections perpendicular to the y-axis are rectangles with a height of 6.
3. Let R be the shaded region in the first quadrant enclosed by the graphs of $y = x$ and $y = 6 - x^2$ and by the y-axis. The region R is the base of a solid. For this solid, each cross section perpendicular to the x-axis is a square. Find the volume of this solid.
4. Let R be the region enclosed by the graph of $y = x^3$ from $x = 0$ to $x = 3$. The base of a solid is the region R. Each cross section of the solid perpendicular to the x-axis is an equilateral triangle. Find the volume of this solid.

(Answers on page 298.)

Arc Length

Finding the length of a straight-line segment can be done using the distance formula since it finds the shortest distance between two points. When asked to find the length of an arc of a continuous function, this is more challenging since the function can be curvy. To find the length of an arc, calculus is required.

The length of an arc, s, can be found by evaluating the following integral:

$$s = \int_a^b \sqrt{1 + (f'(x))^2}\, dx$$

Calculating the integrals involved with finding the arc length might be too difficult to do algebraically. A graphing calculator might be useful in this situation. If using the TI-84 graphing calculator, the following buttons will be used to calculate a definite integral:

MATH 9

Fill in the bounds and integrand, type *x* for *dx*, and then press enter. The calculator will calculate the value of the definite integral.

Finding the arc length of a function is *painless*. It involves the following two steps.

Step 1: Find the derivative of the function.

Step 2: Evaluate the arc length integral.

Example 15:

Find to three decimal places the length of the arc of $f(x) = x^2$ from $x = 1$ to $x = 3$.

Solution:

Step 1: Find the derivative of the function.

The derivative of the function $f(x) = x^2$ is $f'(x) = 2x$.

Step 2: Evaluate the arc length integral:

$$s = \int_1^3 \sqrt{1 + (2x)^2}\, dx$$

Using the TI-84 graphing calculator, $s = 8.268$.

Example 16:

Find the arc length of the curve $y = x^{3/2}$ from (1, 1) to $\left(2, \sqrt{2}^3\right)$ to the nearest tenth.

Solution:

Step 1: Find the derivative of the function.
The derivative of the function $y = x^{3/2}$ is $y' = \frac{3}{2}x^{1/2}$.

Step 2: Evaluate the arc length integral:

$$s = \int_1^2 \sqrt{1 + \left(\frac{3}{2}x^{1/2}\right)^2}\, dx$$

Using the TI-84 graphing calculator, $s = 2.1$.

Example 17:

Let $f(x) = e^{3x}$ and let R be the region in the first quadrant bounded by the graph of f, the coordinate axes, and the vertical line $x = 5$ as shown below.

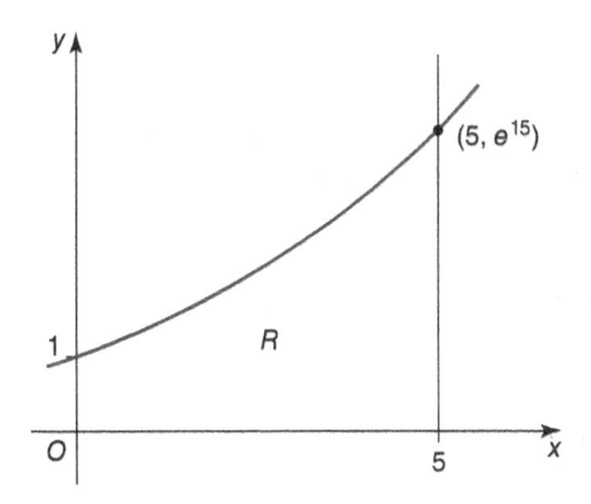

Find the perimeter of R.

Solution:

To find the perimeter of the region, add together the lengths of all the sides of the region. Four sides make up region R. Three are made of straight segments, and one is made by the curve $f(x) = e^{3x}$. The lengths of the straight segments are 1, 5, and e^{15}. To find the length of the curve, the arc length needs to be found.

Step 1: Find the derivative of the function.

The derivative of the function $f(x) = e^{3x}$ is $f'(x) = 3e^{3x}$.

Step 2: Evaluate the arc length integral:

$$s = \int_0^5 \sqrt{1 + (3e^{3x})^2}\, dx$$

Using the TI-84 graphing calculator, $s \approx 3{,}269{,}016.428$.

The perimeter of region $R \approx 1 + 5 + 3{,}269{,}016.428 + e^{15} = 3{,}269{,}022.428 + e^{15}$.

Rectilinear Motion

The previous discussion of rectilinear motion involved only derivatives. With a better understanding of the integral, rectilinear motion can be revisited.

Given the position function, $s(t)$, the velocity and acceleration functions can be found by differentiation: $v(t) = s'(t)$ and $a(t) = v'(t) = s''(t)$.

Given the acceleration function, $a(t)$, the velocity and position functions can be found by antidifferentiation using the indefinite integral: $\int a(t)\, dt = v(t) + C$ and $\int v(t)\, dt = s(t) + C$.

To find the velocity and position functions by integration, more information needs to be given to solve for the constant, C.

Example 18:

Suppose that a particle moves with velocity $v(t) = \cos(\pi t)$ along a coordinate line. If the particle has coordinates $s = 4$ at time $t = 0$, find its position function.

Solution:

To find the position function given the velocity function, find the indefinite integral:

$$\int v(t)dt = s(t) + C$$

$$\int \cos(\pi t)dt = \frac{1}{\pi}\sin(\pi t) + C$$

Since the particle has coordinates $s = 4$ at time $t = 0$, substitute those values into the function and solve for C:

$$s(t) = \frac{1}{\pi}\sin(\pi t) + C$$

$$4 = \frac{1}{\pi}\sin(\pi \bullet 0) + C$$

$$4 = 0 + C$$

$$C = 4$$

Therefore, the position function is $s(t) = \frac{1}{\pi}\sin(\pi t) + 4$.

When utilizing definite integrals, such as integrating velocity over time, the displacement can be found:

$$\int_{t_0}^{t_1} v(t)\,dt = s(t_1) - s(t_0)$$

In contrast, to find the total distance traveled by the particle over the time interval $[t_0, t_1]$ (distance traveled in the positive direction plus the distance traveled in the negative direction), we must integrate the absolute value of the velocity function:

$$\int_{t_0}^{t_1} |v(t)|\,dt = \text{total distance traveled}$$

Example 19:

Suppose that a particle moves on a coordinate line so that its velocity at time t is $v(t) = t^2 - 2t$ m/s.

1. Find the displacement of the particle during the time interval $0 \leq t \leq 3$.
2. Find the distance traveled by the particle during the time interval $0 \leq t \leq 3$.

Solution:

1. Calculate the displacement:

$$\int_0^3 v(t)\,dt = \int_0^3 (t^2 - 2t)\,dt = \left.\frac{t^3}{3} - \frac{2t^2}{2}\right|_0^3$$

$$= \frac{(3)^3}{3} - \frac{2(3)^2}{2} - \left(\frac{(0)^3}{3} - \frac{2(0)^2}{2}\right) = 9 - 9 - 0 + 0 = 0$$

Note: Just because the displacement equals 0 does not mean that the particle did not travel. It is more likely that the particle traveled a certain amount in the positive direction, stopped, turned, and traveled an equal amount in the negative direction.

2. Calculate the distance:

$$\int_0^3 |v(t)|\,dt = \int_0^3 |t^2 - 2t|\,dt$$

To solve this algebraically, the times where the particle changes direction must be found. To do this, set the velocity equal to zero and solve for t:

$$t^2 - 2t = 0$$

$$t(t - 2) = 0$$

$$t = 0,\ t = 2$$

To see the directions the particle moves, create a first derivative sign diagram using velocity.

0 — 2	2 — 3
Choose a t-value in the interval (0, 2)	Choose a t-value in the interval (2, 3)
$t = 1$	$t = 2.5$
Substitute $t = 1$ into the derivative $= v(t)$	Substitute $t = 2.5$ into the derivative $= v(t)$
$v(1) = (1)^2 - 2(1)$	$v(2.5) = (2.5)^2 - 2(2.5)$
$v(1) = -1 < 0$	$v(3) = 1.25 > 0$
Particle is moving left	Particle is moving right

When the particle is moving left, the integral will be a negative value. So, negate the integral from (0, 2):

$$\int_0^3 \left|t^2 - 2t\right| dt = -\int_0^2 (t^2 - 2t)\,dt + \int_2^3 (t^2 - 2t)\,dt$$

$$= -\left(\frac{t^3}{3} - \frac{2t^2}{2}\right)\Bigg|_0^2 + \left(\frac{t^3}{3} - \frac{2t^2}{2}\right)\Bigg|_2^3$$

$$= -\left[\left(\frac{(2)^3}{3} - (2)^2\right) - \left(\frac{(0)^3}{3} - (0)^2\right)\right]$$

$$+ \left[\left(\frac{(3)^3}{3} - (3)^2\right) - \left(\frac{(2)^3}{3} - (2)^2\right)\right]$$

$$= -\frac{8}{3} + 4 + 0 - 0 + 9 - 9 - \frac{8}{3} + 4 = 8 - \frac{16}{3} = \frac{8}{3}$$

Example 20:

The acceleration of a body moving in a straight line is given in terms of time, t, by $a(t) = 8 - 6t$. If the velocity of the body is 25 at $t = 1$, find the displacement of the particle from $t = 4$ to $t = 2$.

Solution:

To find the displacement, the velocity function must first be found:

$$\int a(t)dt = \int (8 - 6t)dt = 8t - \frac{6t^2}{2} + C = 8t - 3t^2 + C = v(t)$$

Since $v = 25$ when $t = 1$, solve for C:

$$25 = 8(1) - 3(1)^2 + C$$

$$C = 20$$

$$v(t) = 8t - 3t^2 + 20$$

Calculate the displacement:

$$\int_2^4 v(t)\,dt = \int_2^4 (8t - 3t^2 + 20)\,dt = \frac{8t^2}{2} - \frac{3t^3}{3} + 20t \Big|_2^4$$

$$= (4(4)^2 - (4)^3 + 20(4)) - (4(2)^2 - (2)^3 + 20(2))$$

$$= 80 - 48 = 32$$

Example 21:

The acceleration of a particle moving along a horizontal line at time t is given by $a(t) = 12t - 2$. If the velocity is 25 when $t = 2$ and the position is 5 when $t = 1$, find the position function, $s(t)$.

Solution:

To find the velocity function, find the indefinite integral of the acceleration function:

$$\int a(t)dt = \int (12t - 2)dt = \frac{12t^2}{2} - 2t + C = 6t^2 - 2t + C = v(t)$$

Since $v(2) = 25$:

$$25 = 6(2)^2 - 2(2) + C \rightarrow C = 5$$

$$v(t) = 6t^2 - 2t + 5$$

To find the position function, find the indefinite integral of the velocity function:

$$\int v(t)dt = \int (6t^2 - 2t + 5)dt = \frac{6t^3}{3} - \frac{2t^2}{2} + 5t + C = s(t)$$

$$s(t) = 2t^3 - t^2 + 5t + C$$

Since $s(1) = 5$:

$$5 = 2(1)^3 - (1)^2 + 5(1) + C \rightarrow C = -1$$

$$s(t) = 2t^3 - t^2 + 5t - 1$$

BRAIN TICKLERS Set # 36

1. Find the length of the arc of $f(x) = \frac{2}{3}x^{3/2}$ from $x = 0$ to $x = 3$.
2. The base of region R in the first quadrant under the graph of $f(x) = 20\sin\left(\frac{\pi x}{30}\right)$ from $0 \leq x \leq 30$ is shown in the figure below. The derivative is $f'(x) = \frac{2\pi}{3}\cos\left(\frac{\pi x}{30}\right)$.

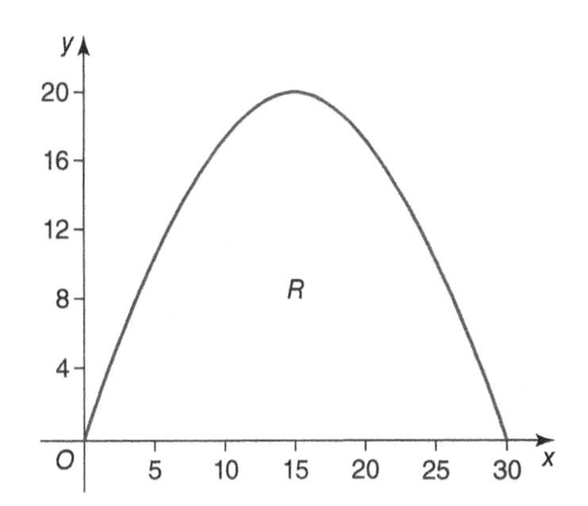

Find the perimeter of region R to four decimal places.

3. A particle moves in a straight line with velocity $v(t) = t^2$. Find the total distance from $t = 1$ to $t = 2$.
4. If the velocity of a particle moving along the x-axis is $v(t) = 2t - 4$ and if at $t = 0$ its position is 4, find the position function, $s(t)$.

(Answers are on page 298.)

BRAIN TICKLERS—THE ANSWERS

Set # 33, page 267

1. 31.5
2. $\frac{1}{6}$
3. 18
4. $\frac{32}{3}$

Set # 34, page 279

1. $\frac{158}{3}\pi$
2. $\frac{1024}{5}\pi$
3. $\frac{5}{14}\pi$
4. 8π

Set # 35, page 289

1. $\frac{81}{2}$
2. 32
3. $\frac{496}{15}$
4. $\frac{2187\sqrt{3}}{28}$

Set # 36, page 297

1. $\frac{14}{3}$
2. 81.8037
3. $\frac{7}{3}$
4. $s(t) = t^2 - 4t + 4$

Chapter 8

Differential Equations

A differential equation is an equation involving a function and one or more of its derivatives.

Understanding Differential Equations

The order of a differential equation is determined by the highest derivative in the equation. For example, if a differential equation has a $\frac{d^2y}{dx^2}$ term and a $\frac{dy}{dx}$ term, the order of the differential equation is 2 because the second derivative is higher than the first derivative.

Example 1:

Determine the order of each differential equation.

1. $\frac{d^2y}{dx^2} + 3\frac{dy}{dx} - 2y = 0$
2. $y' + 2x = y - 3$

Solution:

1. The order is 2.
2. The order is 1.

Separation of Variables

Our goal is to take differential equations of the form $\frac{dy}{dx} = \frac{f(x)}{g(y)}$ and write them into the form $f(x)\ dx = g(y)\ dy$ so that this equation can be integrated. Then it is possible to solve for y as a function of x. This is called the *separation of variables* method.

1+2=3 MATH TALK!

The separation of variables differential equation method can be applied only if the given differential equation, $\frac{dy}{dx}$, is a ratio of a function of x, like $f(x)$, over a function of y, like $g(y)$. Then it is possible to cross multiply so that each variable is on its own side of the equal sign. Next, integrate both sides of the equation where the goal is to solve for y in terms of x.

The separation of variables method is *painless*. It has the following three steps.

Step 1: Cross multiply the ratio so that the variables are separated on opposite sides of the equal sign.

Step 2: Integrate both sides of the equation.

Step 3: Solve for y in terms of x and the constant.

Example 2:

Given $\frac{dy}{dx} = \frac{x}{y}$. Find y, where y is a function of x.

Solution:

Step 1: Cross multiply the ratio so that the variables are separated on opposite sides of the equal sign:

$$\frac{dy}{dx} = \frac{x}{y}$$

$$y \bullet dy = x \bullet dx$$

Step 2: Integrate both sides of the equation:

$$\int y\,dy = \int x\,dx$$

$$\frac{y^2}{2} + C_1 = \frac{x^2}{2} + C_2$$

The subscripts on the constants are just to note that they are different constants for each side.

Step 3: Solve for y in terms of x and the constant.

Start isolating y by first subtracting the constant C_1 from both sides, which leads to a new constant, C_3:

$$\frac{y^2}{2} = \frac{x^2}{2} + C_3$$

Multiply the equation by 2:

$$2\left(\frac{y^2}{2} = \frac{x^2}{2} + C_3\right)$$

$$y^2 = x^2 + C_4$$

Take the square root of both sides:

$$\sqrt{y^2} = \sqrt{x^2 + C_4}$$

$$y = \pm\sqrt{x^2 + C}$$

Example 3:

Find y, where y is a function of x, if $\frac{dy}{dx} = \frac{y}{x^2}$.

Solution:

Step 1: Cross multiply the ratio so that the variables are separated on opposite sides of the equal sign:

$$\frac{dy}{dx} = \frac{y}{x^2}$$

$$x^2 \bullet dy = y \bullet dx$$

$$\frac{1}{y}dy = \frac{1}{x^2}dx$$

Step 2: Integrate both sides of the equation:

$$\int \frac{1}{y}dy = \int \frac{1}{x^2}dx$$

$$\int \frac{1}{y}dy = \int x^{-2}dx$$

$$\ln|y| + C_1 = \frac{x^{-1}}{-1} + C_2$$

Step 3: Solve for y in terms of x and the constant:

$$\ln|y| = -x^{-1} + C_3$$

To isolate y, rewrite the above equation in exponential form:

$$|y| = e^{-x^{-1}+C_3}$$

Use exponent properties:

$$|y| = e^{-x^{-1}} \bullet e^{C_3}$$

Since e^{C_3} is a constant, it is written as C_4:

$$|y| = C_4 e^{-x^{-1}}$$

$$y = \pm C_4 e^{-x^{-1}}$$

The constant C_4 absorbs the sign from the $\pm$ to get the general equation:

$$y = Ce^{-x^{-1}} = Ce^{-1/x}$$

Example 2 and Example 3 found the general solution that corresponds with the given differential equation. This general solution represents infinitely many equations since there is an unknown constant, C. To find a particular solution for a given differential equation, more information needs to be provided, such as a set of coordinates for the particular equation.

Finding a particular solution using the separation of variables method is *painless*. It has the following four steps.

Step 1: Cross multiply the ratio so that the variables are separated on opposite sides of the equal sign.

Step 2: Integrate both sides of the equation.

Step 3: Solve for y in terms of x and the constant.

Step 4: Substitute the given coordinates for x and y to solve for C.

Example 4:

Given $\frac{dy}{dx} = y\sec^2 x$, find a particular solution for y when $y = 5$ and $x = 0$.

Solution:

Step 1: Cross multiply the ratio so that the variables are separated on opposite sides of the equal sign:

$$\frac{dy}{dx} = y\sec^2 x$$

$$dy = y\sec^2 x\,dx$$

$$\frac{1}{y}dy = \sec^2 x\,dx$$

Step 2: Integrate both sides of the equation:

$$\int \frac{1}{y}dy = \int \sec^2 x\,dx$$

$$\ln|y| + C_1 = \tan x + C_2$$

Step 3: Solve for y in terms of x and the constant:

$$\ln|y| = \tan x + C_3$$

$$|y| = e^{\tan x + C_3}$$

$$|y| = C_4 e^{\tan x}$$

$$y = \pm Ce^{\tan x}$$

Step 4: Substitute the given coordinates for x and y to solve for C. Since $y = 5$ and $x = 0$, the equation must be positive since the y-value is positive. Substitute for x and y:

$$(5) = +Ce^{\tan(0)}$$

$$5 = Ce^0$$

$$5 = C(1)$$

$$C = 5$$

Therefore, the particular solution to the given differential equation is $y = 5e^{\tan x}$.

BRAIN TICKLERS Set # 37

1. Given $\frac{dy}{dx} = \frac{y}{x}$, find the general solution for y in terms of x.
2. Find the general solution for y in terms of x of the differential equation $y\,dy = x^2\,dx$.
3. Given the differential equation $\frac{dy}{dx} = \frac{3-x}{y}$, find $f(x)$, the particluar solution with the condition that $f(x) = -4$, when $x = 6$.
4. Find the particular solution to the differential equation $\frac{dy}{dx} = \frac{y}{2\sqrt{x}}$ if $y = 1$ when $x = 4$.

(Answers are on page 330.)

Exponential Growth and Decay

Exponential growth and decay problems have a strong connection to differential equations and real-life situations.

The differential equation $\frac{y'}{y} = k$ means that y' is directly proportional to y since its ratio is equal to the constant, k. In other words, the rate of change of y is proportional to the function y. The general solution to this differential equation is $y = Ce^{kx}$, which is an example of an exponential growth or exponential decay model, depending upon if k is positive or negative.

PAINLESS TIP

The key words to look for to recognize an exponential growth or decay problem are "the rate of change of a quantity is proportional to the quantity." When you come across this phrase or something similar, set up the differential equation. If it is of the form $\frac{y'}{y} = k$, the general solution is automatically $y = Ce^{kx}$. More information will need to be provided to solve for the constants C and k.

Example 5:

The rate of growth of a population of fruit flies is proportional to the number of flies in a jar. If the jar initially contains 20 flies and contains 75 flies on the 3rd day, how many flies will there be on the 7th day?

Solution:

The key phrase, "rate of growth," is found in the first sentence. Let $y =$ number of flies in the jar; then $\frac{y'}{y} = k$ and therefore $y = Ce^{kx}$.

In the last sentence, the given information lists ordered pairs that can be used to substitute into the general solution to solve for C and k. Substitute the first ordered pair, $(0, 20)$, and solve for C:

$$y = Ce^{kx}$$
$$(20) = Ce^{k(0)}$$
$$20 = Ce^{0}$$
$$C = 20$$

The equation is now $y = 20e^{kx}$.

Substitute the second ordered pair, $(3, 75)$, and solve for k:

$$y = 20e^{kx}$$
$$(75) = 20e^{k(3)}$$
$$\frac{75}{20} = e^{3k}$$
$$3k = \ln\left(\frac{15}{4}\right)$$
$$k = \frac{1}{3}\ln\left(\frac{15}{4}\right)$$

The equation is now $y = 20e^{\frac{1}{3}\ln\left(\frac{15}{4}\right)x}$.

To find how many flies there will be on the 7th day, substitute 7 for x:

$$y = 20e^{\frac{1}{3}\ln\left(\frac{15}{4}\right)(7)} \approx 436.9545711$$

There will be 436 whole flies on the 7th day.

Half-life is a common example of exponential decay. A radioactive element's half-life is a unit of time that signifies how long it takes for that radioactive element to lose half of its mass.

Example 6:

If 60 mg of radium are present now and the half-life of radium is 1,690 years, how much radium will be present 1,000 years from now? Round your answer to three decimal places.

Solution:

The key word "half-life" came in the first sentence. Initially, the element has a mass of 60 mg, which can be written as the ordered pair (0, 60). Since radium's half-life is 1,690 years, another ordered pair is (1,690, 30). Using these two ordered pairs, the constants C and k can be found.

Substitute the first ordered pair, (0, 60), and solve for C:

$$\begin{aligned} y &= Ce^{kx} \\ (60) &= Ce^{k(0)} \\ 60 &= Ce^{0} \\ C &= 60 \end{aligned}$$

The equation is now $y = 60e^{kx}$.

Substitute the second ordered pair, (1,690, 30), and solve for k:

$$\begin{aligned} y &= 60e^{kx} \\ (30) &= 60e^{k(1690)} \\ \frac{30}{60} &= e^{1690k} \\ 1690k &= \ln\left(\frac{1}{2}\right) \\ k &= \frac{1}{1690}\ln\left(\frac{1}{2}\right) \end{aligned}$$

The equation is now $y = 60e^{\frac{1}{1690}\ln\left(\frac{1}{2}\right)x}$.

To find how much radium will be present 1,000 years from now, substitute 1,000 for x:

$$y = 60e^{\frac{1}{1690}\ln\left(\frac{1}{2}\right)(1000)} \approx 39.81319148$$

When rounded to three decimal places, there will be 39.813 mg of radium 1,000 years from now.

Logistic Growth

So far, population growth has been demonstrated using exponential growth. However, real-life populations often do not increase forever. There is usually some limiting factor, such as food, living space, or waste disposal. There is a maximum population, or *carrying capacity*, known as M. A more realistic model is the logistic growth model where growth rate is proportional to both the amount present (P) and the carrying capacity that remains ($M - P$).

The logistic differential equation is the following:

$$\frac{dP}{dt} = kP(M - P)$$

In this equation:

$$\frac{dP}{dt} = \text{rate of population growth}$$

$$k = \text{relative growth rate (constant)}$$

$$P = \text{population}$$

$$M = \text{carrying capacity}$$

If the logistic differential equation is solved, the logistic growth model can be found:

$$P = \frac{M}{1 + Ce^{-(Mk)t}}$$

The graph of the logistic growth model tells a lot about the behavior of the logistic differential equation.

Below are the two growth models discussed, exponential and logistic.

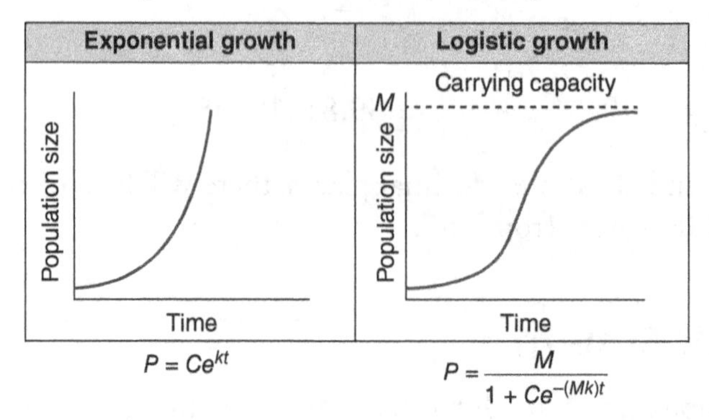

Initially, both graphs look the same. However, at some time, the logistic growth rate changes. It begins to slow down and then eventually levels off when it reaches the carrying capacity, M. The exponential growth graph continues to increase as time increases. This leads to important calculus observations about the two graphs.

Similarities:

1. Both have a similar graph close to time = 0.
2. Both start out growing exponentially.

Differences:

1. For the logistic growth graph, the rate of growth, P', increases in the beginning (because it is concave up, $P'' > 0$) and then decreases (because it is concave down, $P'' < 0$) as the graph approaches its upper limit, M.
2. The logistic growth graph attains a maximum rate when $P = \frac{M}{2}$, which is when the slope of the tangent line is the steepest.
3. The logistic growth graph changes concavity at $P = \frac{M}{2}$.
4. For the logistic growth graph, $\lim_{t \to \infty} P = M$.

Understanding these differences in the two graphs makes it easier to answer logistic growth questions.

Example 7:
Using the logistic differential equation $\frac{dP}{dt} = 0.0002P(1200 - P)$, where t is measured in years, find the following:

1. The carrying capacity of the population
2. The size of the population when it is growing the fastest
3. The rate at which the population is growing when it is growing the fastest

Solution:

Since this is a logistic differential equation, the graph of the logistic growth model and its general formula can be used to help answer the questions:

$$\frac{dP}{dt} = kP(M - P)$$

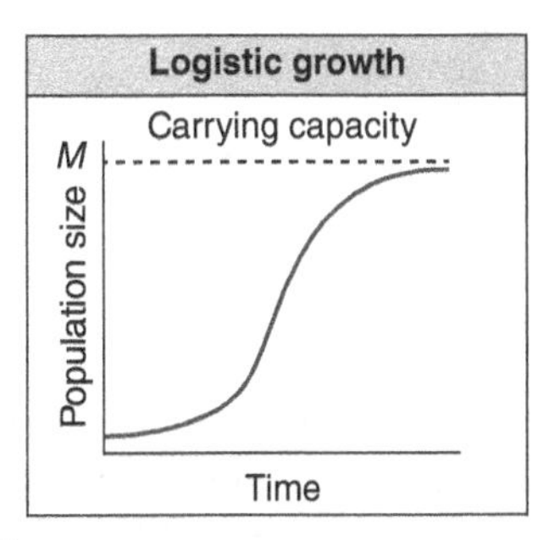

1. The carrying capacity of the population is where the graph levels off and reaches its horizontal asymptote, M. Comparing the general logistic differential equation to the given equation, $M = 1{,}200$. Therefore, the carrying capacity is 1,200.
2. The logistic growth graph attains its greatest rate when $P = \frac{M}{2}$. Therefore, the size of the population when it is growing the fastest $= \frac{1200}{2} = 600$.
3. To find the rate at which the population is growing when it is growing the fastest, substitute $P = 600$ into the logistic differential equation and solve for $\frac{dP}{dt}$:

$$\frac{dP}{dt} = 0.0002(600)(1200 - (600)) = 72$$

Example 8:

Let f be a function such that all points (x, y) on the graph of f satisfy the logistic differential equation $\frac{dy}{dx} = 2y(3 - y)$. Find $\lim_{x \to \infty} f(x)$ and $\lim_{x \to \infty} f'(x)$.

Solution:

Since this is a logistic differential equation, the graph of the logistic growth model and its general formula can be used to help answer the questions.

$$\frac{dP}{dt} = kP(M - P)$$

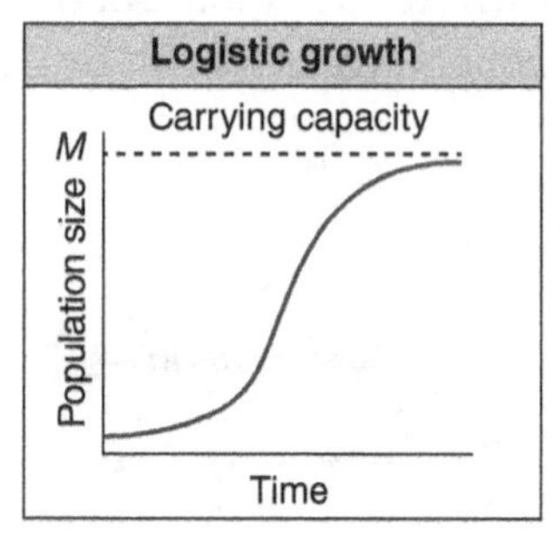

To find $\lim_{x \to \infty} f(x)$, look at the end behavior of the graph. As x approaches infinity, the graph reaches its carrying capacity, M. Comparing the general logistic differential equation to the given logistic differential equation, $M = 3$. Therefore, $\lim_{x \to \infty} f(x) = 3$.

To find $\lim_{x \to \infty} f'(x)$, consider the slopes of the tangent lines of the logistic growth model, which are drawn in below.

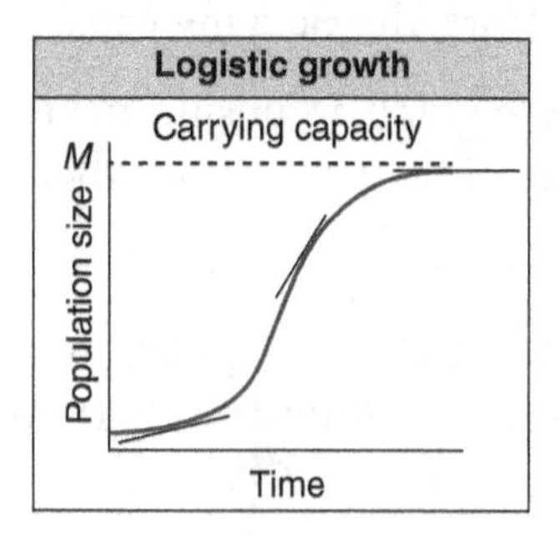

At first, they are positive and increase in steepness. As x approaches infinity, the graph levels off and reaches a horizontal asymptote. The slopes of the tangent lines become zero. Therefore, $\lim_{x \to \infty} f'(x) = 0$.

BRAIN TICKLERS Set # 38

1. Suppose that the value of a certain antique collection increases with age and its rate of appreciation at any time is proportional to its value at that time. If the value of the collection was $15,000 10 years ago and its present value is $28,000, how many years from now will its value be $50,000? Round your answer to two decimal places.
2. The population of a certain town was 25,000 in 1975 and was 34,000 in 1993. If the rate of growth of the population is proportional to the population size, what was the population in the year 1999? Round your answer to three decimal places.
3. The number of squirrels in a town is modeled by the function P that satisfies the logistic differential equation $\frac{dP}{dt} = 0.003P(200 - P)$, where t is the time in years and $P(0) = 50$. What is $\lim_{t \to \infty} P(t)$?
4. Given the graph of the logistic growth model $P(t)$, shown below, what is the size of the population when the rate is growing the fastest?

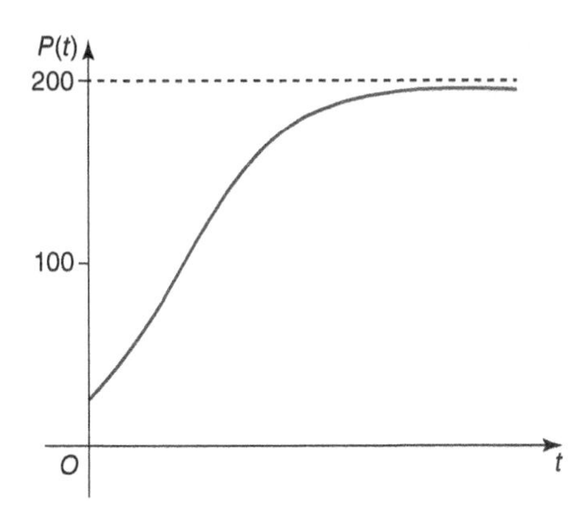

(Answers are on page 330.)

Slope Fields

A slope field is a graphic representation of a differential equation. The graph is comprised of short line segments that represent the value of the derivative at a specific point, (x, y). A slope field is graphed using a grid instead of regular graph paper, as shown below.

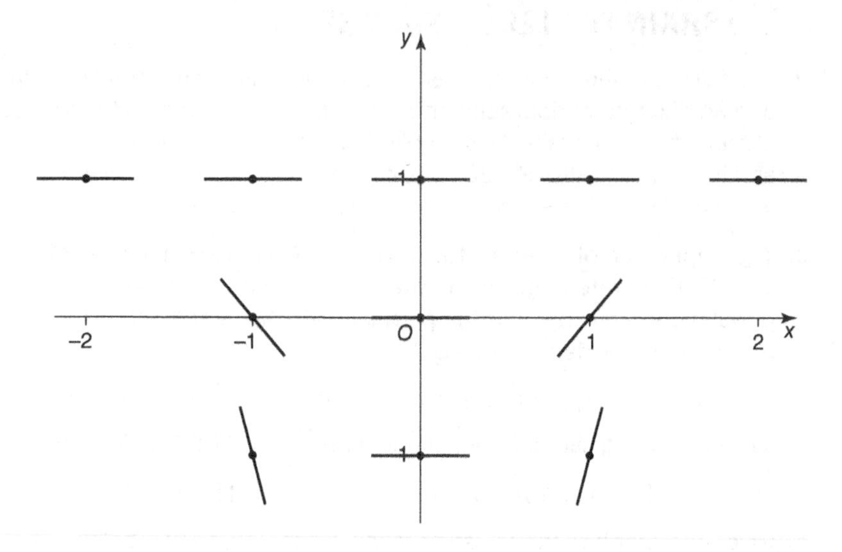

At each grid point is a short line segment drawn to represent the slope of the tangent line at this point. The value of the slope is obtained by substituting the x- and y-values of the grid point into the given differential equation, $\frac{dy}{dx}$. In the figure above, $\frac{dy}{dx} = 0$ along both the y-axis and the horizontal line $y = 1$ since the short line segments are all horizontal, meaning their slopes are 0.

CAUTION—Major Mistake Territory!

A slope field does not represent the graph of the function. It is a graph of short line segments that represent the slopes of the tangent lines at specific coordinates, (x, y). If the path of the line segments is connected, that path will represent the graph of a particular solution to the given differential equation. Technically, a slope field contains infinitely many graphs of particular solutions to a given differential equation. This can be seen below.

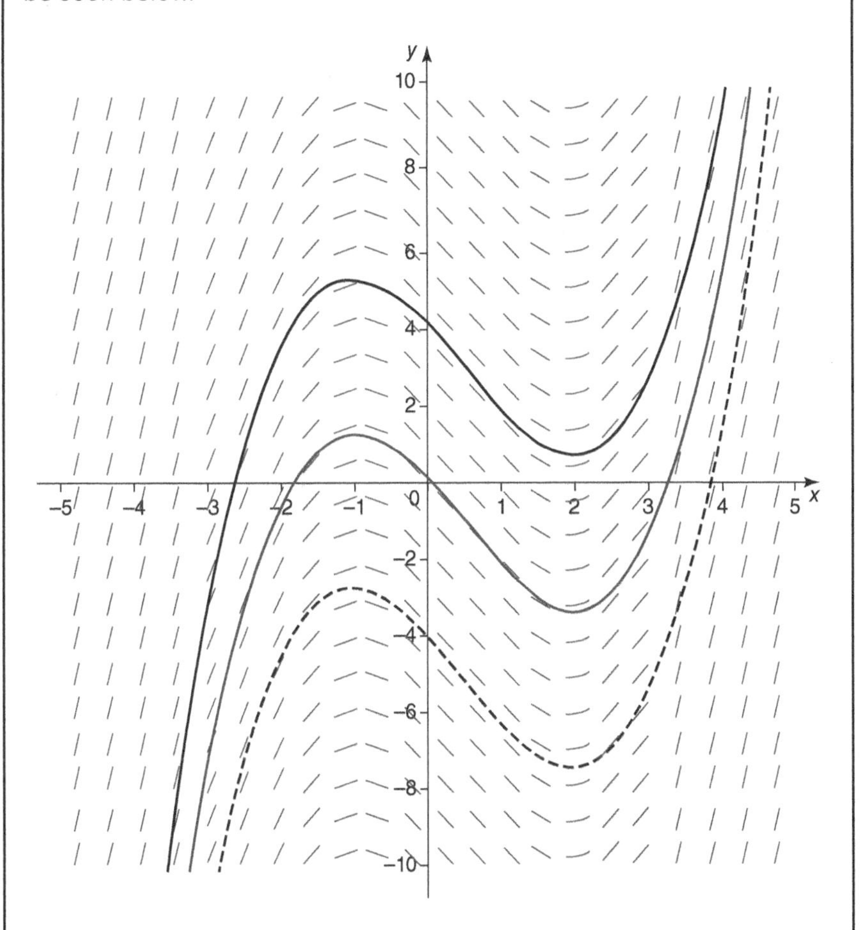

The three different curves represent the graph of three different functions that are particular solutions to a given differential equation. They are all the same shape but are shifted either up or down depending upon certain given corresponding conditions.

Finding the slope field of a given differential equation is *painless.* There are two steps to follow.

Step 1: Fill in a table of values, substituting corresponding x- and y-values into a given differential equation, $\frac{dy}{dx}$.

Step 2: Using a grid, sketch short line segments that represent the slope found from $\frac{dy}{dx}$ at its corresponding grid point (x, y).

Example 9:

Given the differential equation $\frac{dy}{dx} = x$, do the following:

1. Create a slope field given the accompanying grid and table of grid points.
2. Solve the differential equation $\frac{dy}{dx} = x$.

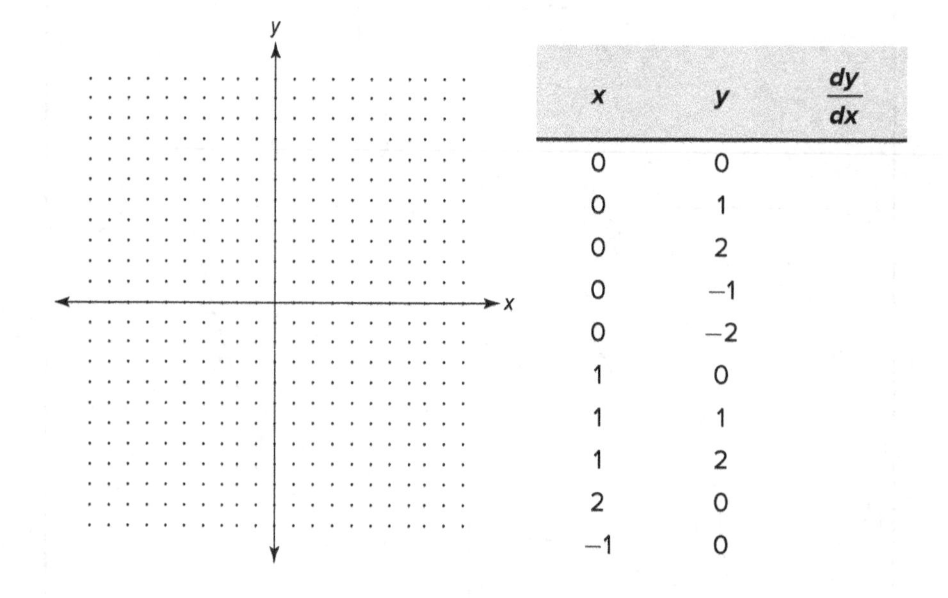

x	y	$\frac{dy}{dx}$
0	0	
0	1	
0	2	
0	−1	
0	−2	
1	0	
1	1	
1	2	
2	0	
−1	0	

Solution:

1.

Step 1: Fill in a table of values, substituting corresponding x- and y-values into a given differential equation, $\frac{dy}{dx}$.

x	y	$\frac{dy}{dx} = x$	
0	0	0	Since $\frac{dy}{dx} = 0$, the slope of the tangent lines at these points is 0. A horizontal line segment should be drawn at these points.
0	1	0	
0	2	0	
0	–1	0	
0	–2	0	
1	0	1	Since $\frac{dy}{dx} = 1$, the slope of the tangent lines at these points is 1. The same positive sloped line segment should be drawn at these points.
1	1	1	
1	2	1	
2	0	2	A steeper positive sloped line segment should be drawn here.
–1	0	–1	A negative sloped line segment with the same steepness as 1 should be drawn here.

Step 2: Using a grid, sketch short line segments that represent the slope found from $\frac{dy}{dx}$ at its corresponding grid point (x, y).

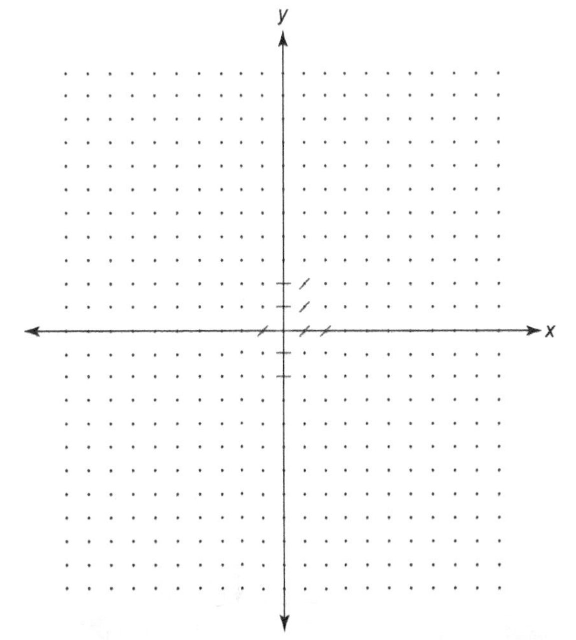

2.

Step 1: Cross multiply the ratio so that the variables are separated on opposite sides of the equal sign.

$$\frac{dy}{dx} = x$$

$$dy = x\,dx$$

Step 2: Integrate both sides of the equation:

$$\int dy = \int x\,dx$$

$$y + C_1 = \frac{x^2}{2} + C_2$$

Step 3: Solve for y in terms of x and the constant:

$$y = \frac{x^2}{2} + C$$

1+2=3 MATH TALK!

There are a few interesting notes about Example 9. One note is that for the points with the same x-coordinate, all of the short line segments are the same because they all have the same slope. This occurs because the differential equation is based upon only the x-value. Recognizing this allows you to draw in more short line segments to get a more comprehensive slope field.

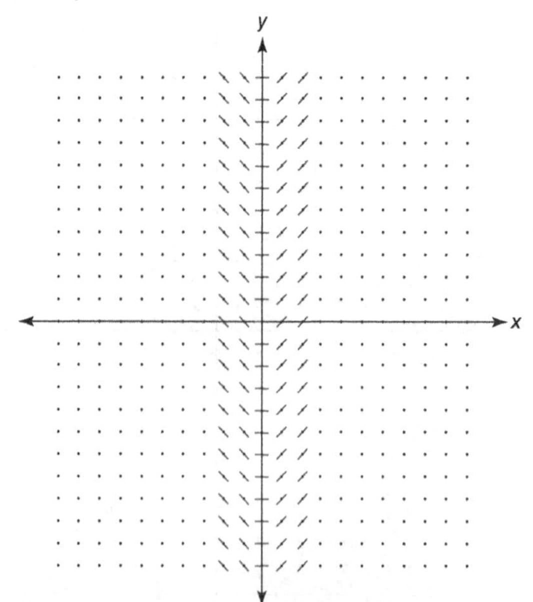

It is also important to note that this slope field represents the graphs of infinitely many particular solutions. They all have the same general shape but are just shifted up or down depending on particular points through which they pass.

The general solution, $y = \frac{x^2}{2} + C$, is a quadratic whose graph is a parabola. You can see that the overall shape that can be traced along the slope field is a parabola. The reason why there are infinitely many graphs is because of C, the constant.

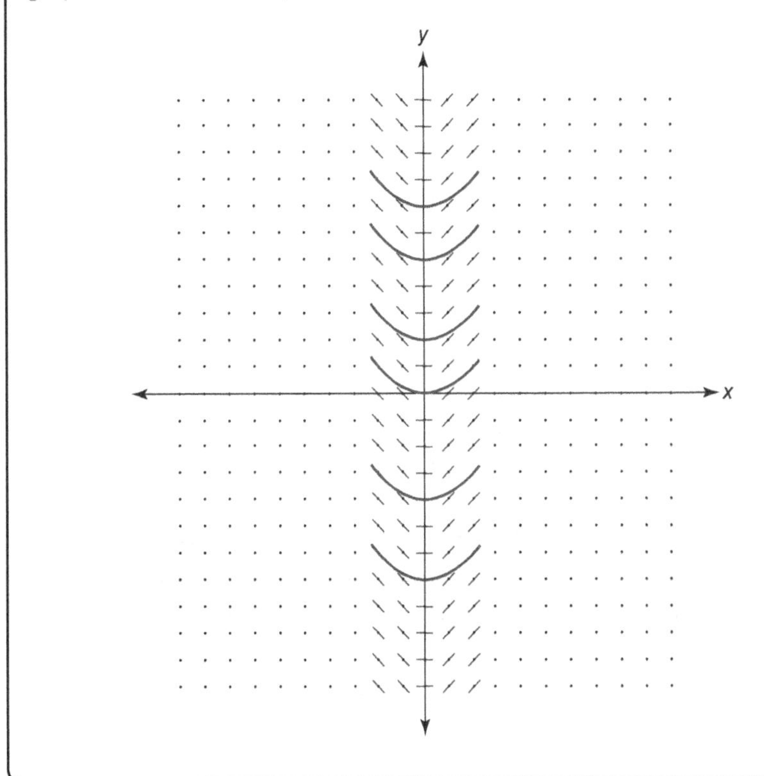

A slope field can be used to identify the graphs of the general solutions of a differential equation. A slope field can also be used to find the graph of a particular solution. If given a starting point and asked to graph a particular solution on a slope field, begin at the starting point and connect the short line segments so that the graph is parallel to the slope lines and is an average of the slopes if it goes between lines. Go both to the right and left of the starting point.

Example 10:

The diagram below shows the slope field for the differential equation $\frac{dy}{dx} = -\frac{0.4x}{y}$.

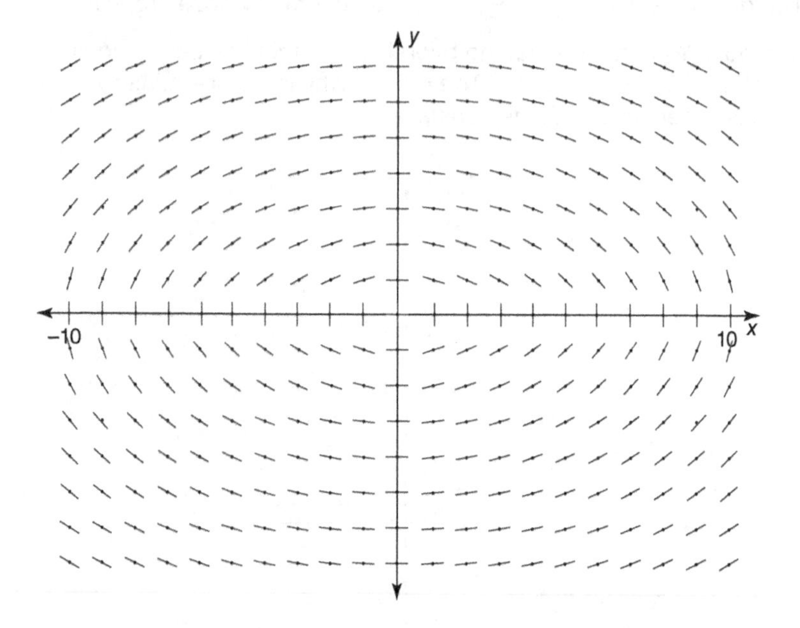

1. Starting at the point (0, 6), draw a graph representing the particular solution of the differential equation that contains that point. What geometrical figure does the graph seem to be?
2. Find the general solution to the differential equation $\frac{dy}{dx} = -\frac{0.4x}{y}$.

Solution:

1. Starting at the point (0, 6), trace along the short line segments.

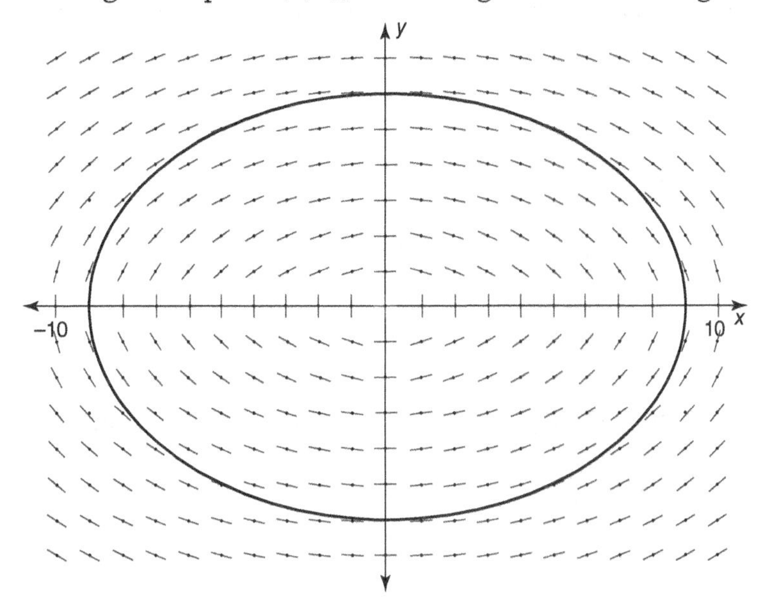

The graph appears to be an ellipse.

2.

Step 1: Cross multiply the ratio so that the variables are separated on opposite sides of the equal sign:

$$\frac{dy}{dx} = -\frac{0.4x}{y}$$

$$y\,dy = -0.4x\,dx$$

Step 2: Integrate both sides of the equation:

$$\int y\,dy = \int -0.4x\,dx$$

$$\frac{y^2}{2} + C_1 = \frac{-0.4x^2}{2} + C_2$$

$$\frac{y^2}{2} + C_1 = \frac{-x^2}{5} + C_2$$

Step 3: Solve for y in terms of x and the constant.

Since the graph of the particular solution appears to be an ellipse, write the equation in the general form of an ellipse, $\frac{x^2}{A}+\frac{y^2}{B}=C$:

$$\frac{y^2}{2}=\frac{-x^2}{5}+C$$

Bring the variables to the left side of the equation:

$$\frac{x^2}{5}+\frac{y^2}{2}=C$$

This follows the general equation of an ellipse centered at the origin, $\frac{x^2}{A}+\frac{y^2}{B}=C$.

A few strategies may be helpful when matching differential equations with their slope fields.

Strategy 1: If given $y'=f(x)$, integrate $f(x)$ to find y, provided $f(x)$ is easy to integrate.

Strategy 2: If the slopes are constant along all vertical lines of a slope field, x is the only component of the differential equation.

Strategy 3: If the slopes are constant along all horizontal lines of a slope field, y is the only component of the differential equation.

Strategy 4: Examine the slope field at specific (x, y) values.

Example 11:

Match the slope fields with their differential equations.

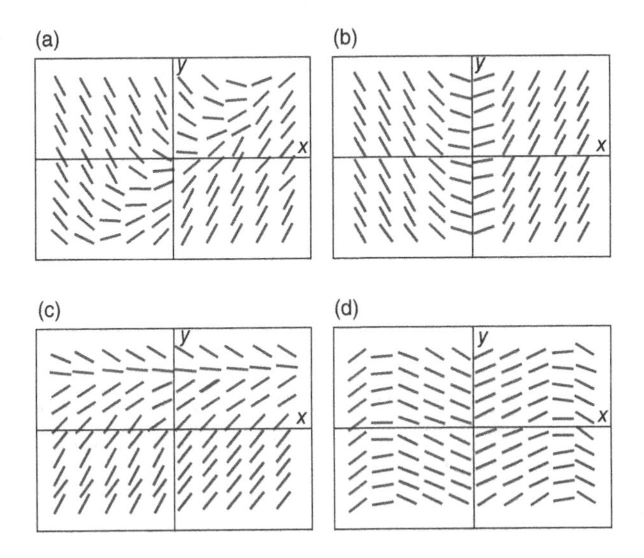

1. $\dfrac{dy}{dx} = x$
2. $\dfrac{dy}{dx} = 2 - y$
3. $\dfrac{dy}{dx} = \sin x$
4. $\dfrac{dy}{dx} = x - y$

Solution:

1. (b) Since the slopes of the graph are constant along all vertical lines, its differential equation can have only the variable x.
2. (c) Since the slopes of the graph are constant along all horizontal lines, its differential equation can only have the variable y.
3. (d) Since the slopes of the graph are constant along all vertical lines, its differential equation can only have the variable x. Also, when connecting the short line segments, the graph appears to resemble the sine curve.

4. (a) Since the slopes of the graph were not constant along all vertical and horizontal lines, the differential equation must have both the variables x and y. Substituting different corresponding x- and y-values into the differential equation also matches the positive and negative corresponding slopes of the graph.

Euler's Method

The previous topic showed how to find solution curves to first-order differential equations graphically using slope fields. Once these curves are found, other points can be discovered using the solution curves.

Euler's method is an alternative algebraic way to find points on solution curves. To use Euler's method, you will be given a starting point on a curve, a differential equation, and a Δx. You are to use the givens to approximate the coordinates of another point on the curve.

The differential equation is used to find the corresponding Δy. So, you can begin at the starting point and use the Δx and Δy to get closer and closer to the endpoint you are approximating. Since you substitute the new x- and y-coordinates into the differential equation each time, the Δy gets closer to the curve and your approximation becomes better.

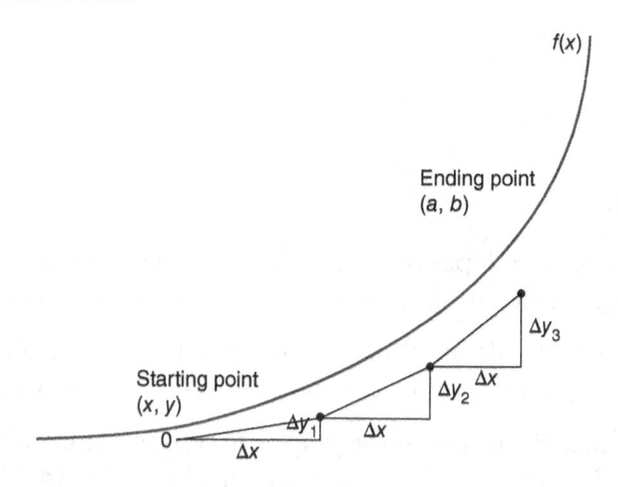

1+2=3 MATH TALK!

This may remind you of using the tangent line to approximate the coordinates of a point on a curve.

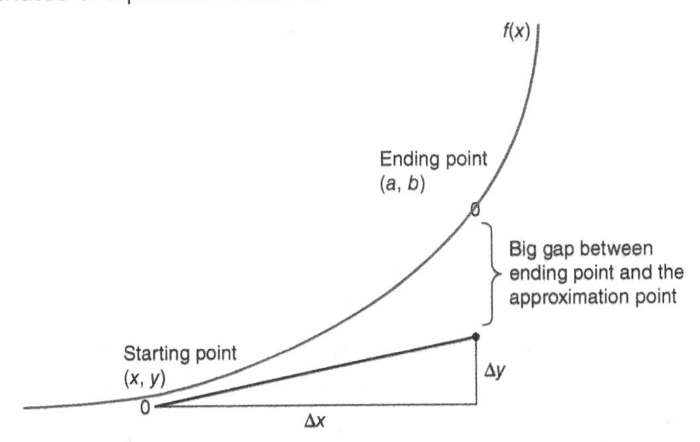

Using one tangent line to approximate the coordinates of a point to a curve can often leave a big gap between the actual coordinates of the ending point and the approximation coordinates. That occurs because the calculations use just the slope of the tangent line at that one starting point.

Euler's method runs through multiple steps so that the approximation is a better fit to the curve, as shown below.

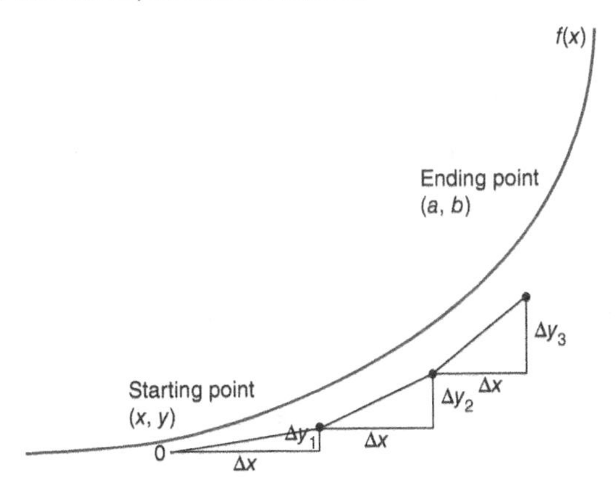

The reason why the Δx stays the same in the diagram above is because that value will be given to you and is a set value. The Δy changes each time because you find a new Δy using the differential equation each time you find the next set of coordinates. Having a new Δy allows the slopes to get closer to the curve every time, and your approximation of the endpoint will be better than if you found just one Δy and used it the entire time.

Using Euler's method to approximate the function value of a curve is *painless*. It follows five steps.

Step 1: Create a table using the headings shown below.

(x, y)	$\Delta y = \frac{dy}{dx} \bullet \Delta x$	$(x + \Delta x,\ y + \Delta y)$

Step 2: Begin filling in the table by writing the coordinates of the starting point in the first column.

Step 3: Calculate Δy by substituting the x- and y-values into the given differential equation and multiplying by the given Δx.

Step 4: Find the new coordinates by adding the given Δx to x and the calculated Δy to y.

Step 5: Go back to the first column, and fill in the new coordinates. Repeat Steps 3 and 4 until you have arrived at the ending point.

Example 12:

If $\frac{dy}{dx} = xy^2$ and $y = -1$ when $x = 2$, what is the approximate value of y when $x = 2.4$ using Euler's method with $\Delta x = 0.2$?

Solution:

Step 1: Create a table using the headings shown below.

(x, y)	$\Delta y = \frac{dy}{dx} \bullet \Delta x = xy^2 \bullet 0.2$	$(x + \Delta x,\ y + \Delta y)$

Step 2: Begin filling in the table by writing the coordinates of the starting point in the first column.

(x, y)	$\Delta y = \frac{dy}{dx} \bullet \Delta x = xy^2 \bullet 0.2$	$(x + \Delta x, y + \Delta y)$
$(2, -1)$		

Step 3: Calculate Δy by substituting the x- and y-values into the given differential equation and multiplying by the given Δx.

(x, y)	$\Delta y = \frac{dy}{dx} \bullet \Delta x = xy^2 \bullet 0.2$	$(x + \Delta x, y + \Delta y)$
$(2, -1)$	$\Delta y = (2)(-1)^2 \bullet 0.2 = 0.4$	

Step 4: Find the new coordinates by adding the given Δx to x and the calculated Δy to y.

(x, y)	$\Delta y = \frac{dy}{dx} \bullet \Delta x = xy^2 \bullet 0.2$	$(x + \Delta x, y + \Delta y)$
$(2, -1)$	$\Delta y = (2)(-1)^2 \bullet 0.2 = 0.4$	$(2 + 0.2, -1 + 0.4)$ $= (2.2, -0.6)$

Step 5: Go back to the first column, and fill in the new coordinates. Repeat Steps 3 and 4 until you have arrived at the ending point.

(x, y)	$\Delta y = \frac{dy}{dx} \bullet \Delta x = xy^2 \bullet 0.2$	$(x + \Delta x, y + \Delta y)$
$(2, -1)$	$\Delta y = (2)(-1)^2 \bullet 0.2 = 0.4$	$(2 + 0.2, -1 + 0.4)$ $= (2.2, -0.6)$
$(2.2, -0.6)$	$\Delta y = (2.2)(-0.6)^2 \bullet 0.2$ $= 0.1584$	$(2.2 + 0.2, -0.6 + 0.1584)$ $= (2.4, -0.4416)$

Using Euler's method, when $x = 2.4$, $y \approx -0.4416$.

CAUTION—Major Mistake Territory!

Euler's method has many steps, so it is important to take your time and stay organized! Do not rush. When using Euler's method and repeating the steps a second time, students often forget to substitute the new *x*- and *y*-values or forget to multiply the differential equation by Δx. Be sure to go slowly and check all of your work before proceeding to the next step.

Example 13:

The solution of the differential equation $\frac{dy}{dx} = -\frac{x^2}{y}$ contains the point $(3, -2)$. Use Euler's method with $\Delta x = -0.3$ to approximate y when $x = 2.7$.

Solution:

Step 1: Create a table using the headings shown below.

(x, y)	$\Delta y = \frac{dy}{dx} \bullet \Delta x = -\frac{x^2}{y} \bullet -0.3$	$(x + \Delta x,\ y + \Delta y)$

Step 2: Begin filling in the table by writing the coordinates of the starting point in the first column.

(x, y)	$\Delta y = \frac{dy}{dx} \bullet \Delta x = -\frac{x^2}{y} \bullet -0.3$	$(x + \Delta x,\ y + \Delta y)$
$(3, -2)$		

Step 3: Calculate Δy by substituting the x- and y-values into the given differential equation and multiplying by the given Δx.

(x, y)	$\Delta y = \frac{dy}{dx} \bullet \Delta x = -\frac{x^2}{y} \bullet -0.3$	$(x + \Delta x,\ y + \Delta y)$
$(3, -2)$	$\Delta y = -\frac{(3)^2}{-2} \bullet -0.3 = -1.35$	

Step 4: Find the new coordinates by adding the given Δx to x and the calculated Δy to y.

(x, y)	$\Delta y = \frac{dy}{dx} \bullet \Delta x = -\frac{x^2}{y} \bullet -0.3$	$(x + \Delta x,\ y + \Delta y)$
$(3, -2)$	$\Delta y = -\frac{(3)^2}{-2} \bullet -0.3 = -1.35$	$(3 + -0.3, -2 + -1.35)$ $= (2.7, -3.35)$

Step 5: Go back to the first column, and fill in the new coordinates. Repeat Steps 3 and 4 until you have arrived at the ending point.

(x, y)	$\Delta y = \frac{dy}{dx} \bullet \Delta x = -\frac{x^2}{y} \bullet -0.3$	$(x + \Delta x,\ y + \Delta y)$
$(3, -2)$	$\Delta y = -\frac{(3)^2}{-2} \bullet -0.3 = -1.35$	$(3 + -0.3, -2 + -1.35)$ $= (2.7, -3.35)$
$(2.7, -3.35)$		

Using Euler's method, when $x = 2.7$, $y \approx -3.35$.

1+2=3 MATH TALK!

There are a few ways to approximate the function value for a given curve and a given *x*-value, such as using the equation of a tangent line or Euler's method. Remember that these are just approximations. They will be close to the actual value, and sometimes one method is better than the other. The best way to find the actual value of a function is to find the particular solution of a differential equation by integrating and then substituting the *x*-value into the equation. Then you are guaranteed to have the corresponding *y*-value.

BRAIN TICKLERS Set # 39

1. Sketch a slope field for the given differential equation, $\frac{dy}{dx} = \frac{xy}{2}$, at the nine points specified on the grid below.

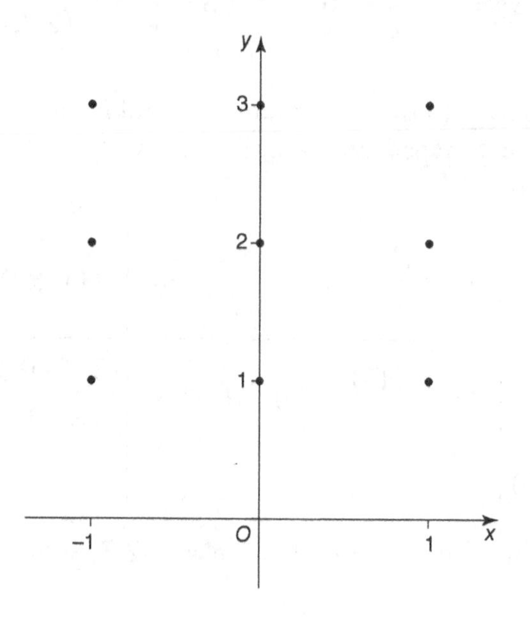

2. Given the differential equation $\frac{dy}{dx} = \frac{-x}{y}$, create a slope field given the accompanying grid. Draw in the graph of a particular solution that passes through the point (0, –1).

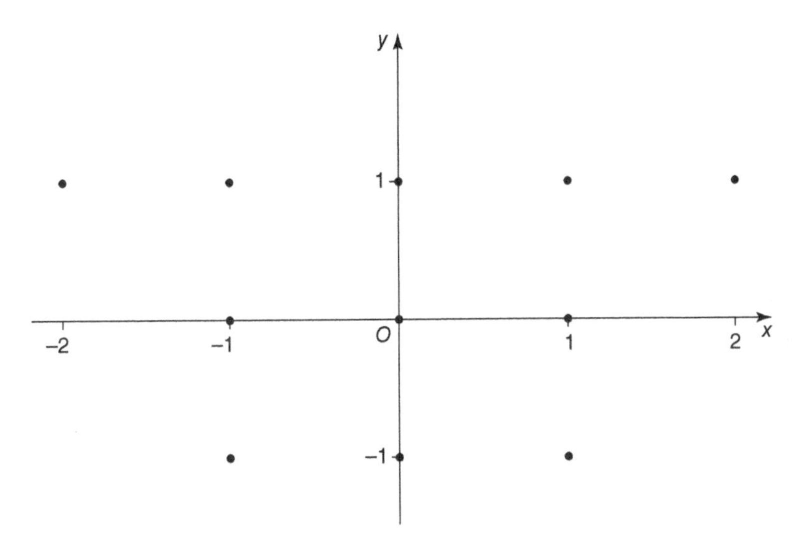

3. If $\frac{dy}{dx} = \frac{x - y}{2y}$ and $y = -2$ when $x = 3$, find the approximate value of y when $x = 3.2$ and $\Delta x = 0.2$ using Euler's method.

4. If $\frac{dy}{dx} = \frac{y}{8}(6 - y)$ and $y = 8$ when $x = 0$, find the approximate value of y when $x = 1$ and $\Delta x = 0.5$ using Euler's method.

(Answers are on pages 330–331.)

BRAIN TICKLERS—THE ANSWERS

Set # 37, page 304

1. $y = C|x|$
2. $y = \pm\sqrt{\frac{2x^3}{3} + C}$
3. $f(x) = y = -\sqrt{6x - x^2 + 16}$
4. $y = \frac{1}{e^2}e^{\sqrt{x}} = e^{-2}e^{\sqrt{x}} = e^{\sqrt{x}-2}$

Set # 38, page 311

1. 9.29
2. 37,669.676
3. 200
4. 100

Set # 39, pages 328-329

1.

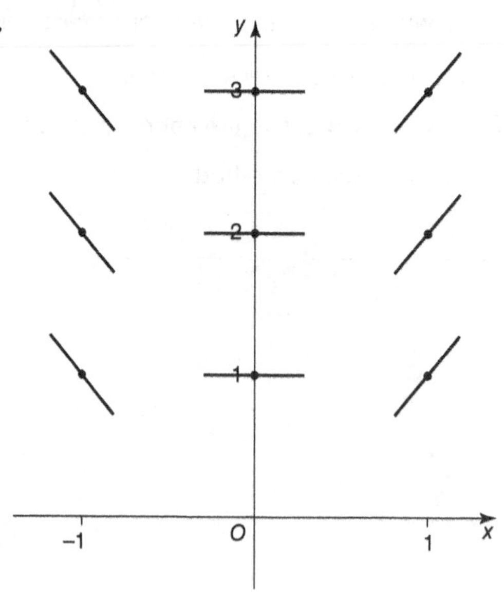

2.

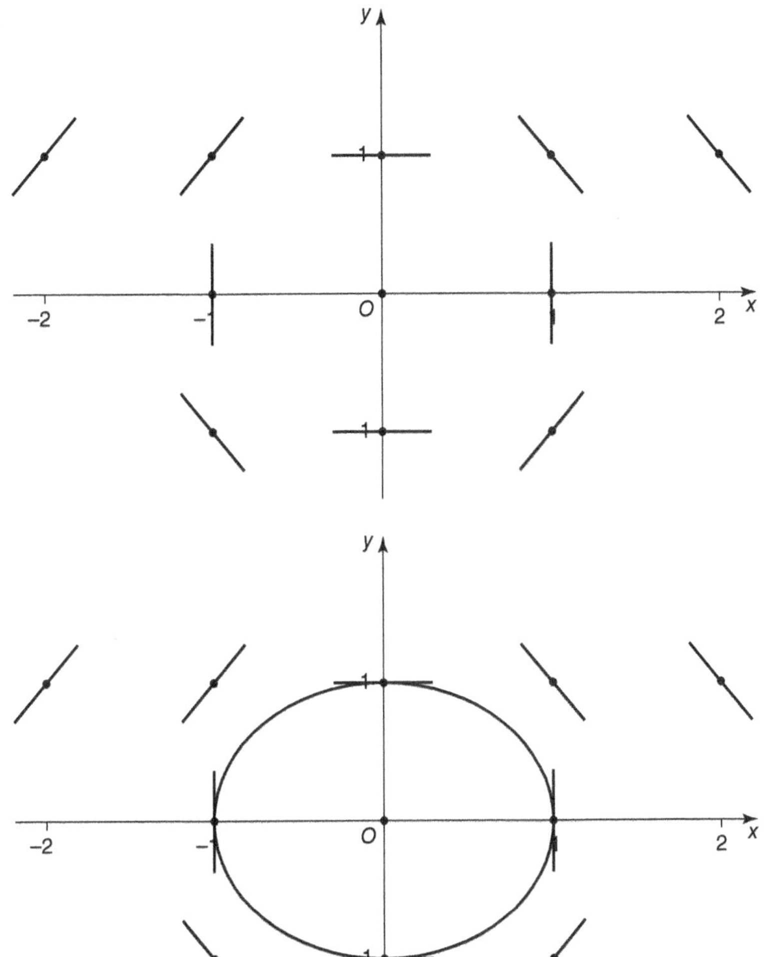

3. −2.25

4. 6.5625

Appendix

Important Derivatives, Formulas, Integrals, Limits, and Theorems

Important Derivatives

Chain Rule for Derivatives:

If $h(x) = g(f(x))$, then $h'(x) = g'(f(x)) \cdot f'(x)$.

Exponential Derivatives:

$$\frac{d}{dx}(b^x) = b^x \ln b$$

$$\frac{d}{dx}(e^x) = e^x$$

Inverse Function Derivative:

$$(f^{-1}(x))' = \frac{1}{f'(f^{-1}(x))}$$

Inverse Trigonometric Derivatives:

$$\frac{d}{dx}(\sin^{-1} x) = \frac{1}{\sqrt{1 - x^2}}$$

$$\frac{d}{dx}(\cos^{-1} x) = \frac{-1}{\sqrt{1 - x^2}}$$

$$\frac{d}{dx}(\tan^{-1} x) = \frac{1}{x^2 + 1}$$

$$\frac{d}{dx}(\cot^{-1} x) = \frac{-1}{x^2 + 1}$$

$$\frac{d}{dx}(\sec^{-1} x) = \frac{1}{|x|\sqrt{x^2 - 1}}$$

$$\frac{d}{dx}(\csc^{-1} x) = \frac{-1}{|x|\sqrt{x^2 - 1}}$$

Logarithmic Derivatives:

$$\frac{d}{dx}(\log_b x) = \frac{1}{x \ln b}$$

$$\frac{d}{dx}(\ln x) = \frac{1}{x}$$

Power Rule for Derivatives:

If $f(x) = ax^n$, then $f'(x) = n \cdot ax^{n-1}$.

Product Rule for Derivatives:

If $h(x) = f(x) \cdot g(x)$, then $h'(x) = f(x) \cdot g'(x) + g(x) \cdot f'(x)$.

Quotient Rule for Derivatives:

If $h(x) = \frac{f(x)}{g(x)}$, then $h'(x) = \frac{g(x) \cdot f'(x) - f(x) \cdot g'(x)}{(g(x))^2}$.

Trigonometric Derivatives:

$$\frac{d}{dx}(\sin x) = \cos x$$

$$\frac{d}{dx}(\cos x) = -\sin x$$

$$\frac{d}{dx}(\tan x) = \sec^2 x$$

$$\frac{d}{dx}(\cot x) = -\csc^2 x$$

$$\frac{d}{dx}(\sec x) = \tan x \sec x$$

$$\frac{d}{dx}(\csc x) = -\cot x \csc x$$

Important Formulas

Arc Length:

$$s = \int_a^b \sqrt{1 + (f'(x))^2}\, dx$$

Area Under a Curve:

$$A = \int_a^b f(x)\, dx$$

Average Function Value:

$$f(c) = \frac{1}{b - a} \int_a^b f(x)\, dx$$

Logistic Differential Equation:

$$\frac{dP}{dt} = kP(M - P)$$

Volume of a Solid with a Known Cross Section:

$V = \int_a^b A(x)\, dx$, $A(x)$ is the area function of the cross section.

Volume of a Solid of Revolution:

$$V = \pi \int_a^b (f(x))^2 dx$$

Important Integrals

Exponential Integrals:

$$\int b^x dx = \frac{b^x}{\ln b} + C$$

$$\int e^x dx = e^x + C$$

Integration by Parts:

$$\int u\, dv = uv - \int v\, du$$

Inverse Trigonometric Integrals:

$$\int \frac{dx}{\sqrt{1 - x^2}} = \sin^{-1} x + C$$

$$\int \frac{dx}{1 + x^2} = \tan^{-1} x + C$$

$$\int \frac{dx}{x\sqrt{x^2 - 1}} = \sec^{-1}|x| + C$$

Logarithmic Integrals:

$$\int \frac{1}{x}\,dx = \ln|x| + C$$

Power Rule for Integrals:

If $f(x) = ax^n$, then $\int ax^n dx = \dfrac{ax^{n+1}}{n+1} + C$.

Trigonometric Integrals:

$$\int \sin x\,dx = -\cos x + C$$

$$\int \cos x\,dx = \sin x + C$$

$$\int \tan x\,dx = \ln|\sec x| + C$$

$$\int \cot x\,dx = \ln|\sin x| + C$$

$$\int \sec^2 x\,dx = \tan x + C$$

$$\int \csc^2 x\,dx = -\cot x + C$$

$$\int \sec x \tan x\,dx = \sec x + C$$

$$\int \csc x \cot x\,dx = -\csc x + C$$

Important Limits

$$\lim_{x\to 0} \frac{\sin x}{x} = 1$$

$$\lim_{x\to 0} \frac{\cos x - 1}{x} = \lim_{x\to 0} \frac{1 - \cos x}{x} = 0$$

Important Theorems

Extreme Value Theorem:

If f is a continuous function on a closed interval $[a, b]$, then f has both a maximum and a minimum value on $[a, b]$.

Intermediate Value Theorem:

If f is continuous on a closed interval $[a, b]$ and C is any number between $f(a)$ and $f(b)$, inclusive, there is at least one number x in the interval $[a, b]$ such that $f(x) = C$.

Mean Value Theorem for Derivatives:
If f is a continuous function on the closed interval $[a, b]$ and differentiable on the open interval (a, b), there is at least one point c in (a, b) such that $f'(c) = \frac{f(b) - f(a)}{b - a}$.

Mean Value Theorem for Integrals:
If f is a continuous function over $[a, b]$, there is at least one point c in $[a, b]$ such that $\int_a^b f(x)\,dx = (b - a) \cdot f(c)$.

Index